# Flickering Clusters

# Flickering Clusters

## Women, Science, and Collaborative Transformations

Cheryl Ney

Jacqueline Ross

Laura Stempel

ISBN 0-9679587-0-9

Published by University Communications for
the University of Wisconsin System Women's Studies Consortium
1660 Van Hise Hall
1220 Linden Drive
Madison, WI 53706

Distributed by the University of Wisconsin Press, Madison, Wisconsin
www.wisc.edu/wisconsinpress/

# Contents

# Foreword

**Sheila Tobias***

Three advances in the search for ways of increasing the numbers and the success rate of women in science come together in this volume. The first is the acknowledgment—not publicly accepted until the late 1970s—that female underenrollment in college mathematics and science majors is a *feminist issue*. In the critique of the curriculum which engaged the founders of the field that came to be known as "women's studies" from 1968 on, the omissions, distortions, and trivialization[1] in the humanities and social sciences were so egregious (and familiar) as to demand full attention.

In the late 1970s, however, my own and others' work on "math avoidance" in women and girls,[2] together with the emergence of women's caucuses, status-of-women commissions, and task forces within physics, chemistry, the life sciences, and the continuing presence of groups like the Society for Women Engineers (linked to the Wisconsin project in the presence of Ethel Sloane), extended that curriculum critique to science/mathematics textbooks, teaching styles, and to what became known as the "classroom culture" itself.

By the time the venerable American Association of University Women produced their comprehensive report on "How Our Schools Shortchange Girls" in the early 1990s, educators knew that the "chilly climate" for women and girls was as much if not more of a barrier to achievement in the study of mathematics, science, and engineering, as it was in other disciplines. Some feminists were even extending their critique to the "androcentric" style, methods, and selection of topics that drives science research itself.

The second advance in the evolution of the issue represented in the Wisconsin project is the shift in *strategy* from a focus on elementary and secondary schooling to a concern with introductory first-year courses in college, known colloquially and functioning operationally as "weed-out" courses. Data collected by the NSF and widely publicized by a number of researchers[3] revealed that (apart from pre-meds) only some 20 percent of those who enter college intending to major in the sciences, mathematics, or a technical subject, actually complete that major, the others leaving sometime between their freshman and junior years.

This exodus from the sciences and mathematics[4] is not unique to female students, but they constitute a disproportionate number of those who are weeded out. Hence, first-year courses deserved at least as much reform and improvement as middle and high-school science. Or so some of us thought.

But what kind of reform? And how to intervene at college, where faculty individually control the classroom and collectively the curriculum? It is one thing for the National Council of Teachers of Mathematics and the National Research Council to promulgate "national standards" for elementary and secondary mathematics and science (respectively); quite another to tell college and university faculty what to teach and how to teach their students.

And so the third advance in the search for a solution—finding a workable and long-lasting intervention—waited to be discovered. At first, it was thought that some "magic bullet" in the form of a near-perfect curriculum and better teaching strategies would compensate young women for the "chilly climate"; or that special sections, such as "math-anxiety reduction," would keep them on track.

But, as American Association for the Advancement of Science (AAAS) researchers Marsha Matyas and Shirley Malcolm documented in their comprehensive review of compensatory programs,[5] special programs rarely survive the cessation of funding and little is to be seen of their efforts some few years later. How, then, to effect systemic change at the college and university levels? How to attract and *retain* women (and minority) students in science, mathematics, and engineering?

The program detailed in this volume goes far to answer these questions and to move us from a policy of short-term interventions to *systemic reform*. As Daryl Chubin writes in *Implementing Science Education Reform: Are We Making an Impact*,[6] "systemic reform is more of a transfusion—initially of resources, but moreover of spirit and vision—that propels continuous (internal) improvement *without* continuous (external) incentives"—an apt description of the Women and Science Program at the University of Wisconsin.

The Wisconsin project was designed to be Systemwide and lasting in its effects by changing the way entire faculties think about recruitment and retention in their disciplines. In place of short-term special programs, the Wisconsin team invented a new strategy: the Distinguished Visiting Professor who, in conjunction with local Faculty Fellows, would bring to a host institution the new consciousness of what women students want and need, together with appropriate and research-based teaching and curriculum innovations.

Basic to the new strategy is a deeper understanding of the ideology of recruitment and advancement in science, mathematics, and engineering. In place of "cloning"—the search among out groups for in-group types—the project offers faculty the opportunity to reflect on their teaching and to apply the findings of the considerable amount of research in feminist pedagogy that has hitherto not found its way into science teaching.

Distinguished Visiting Professors give professional development workshops for faculty and staff, design and teach model courses, initiate one-on-one discussions with faculty, and sit in on departmental meetings. Considered experts in meeting the needs of underrepresented students, these visitors act as "change agents" in the profoundest sense: changing attitudes, habits, and traditions in the departments they visit, leaving a core faculty in their wake committed to carrying out and carrying on the process.

The distinguishing feature of the University of Wisconsin program is process. Because systemic reform is new and we don't know what it will take to succeed, it is inappropriate to demand results in any particular time frame. This kind of change might have to include years of talking around the issue to really make permanent inroads into departments that—even after the University of Wisconsin Chancellors and provosts on every campus gave this project their imprimatur—still were resistant to change. Thus, it is not surprising that throughout this volume, the refrain "much more needs to be done" recurs. Nonetheless, faculty and administrators across the UW System are committed to carrying on the dialogue, and the success of the Women and Science Program's ongoing Summer Institutes demonstrates commitment and interest that extends beyond Wisconsin.

The reader will find much to savor in this useful, interesting, and important volume. In addition to a literature review, the voices of participants are almost audible as we read personal essays from Distinguished Visiting Professors and faculty from host institutions reflecting on the process of change.

* Sheila Tobias, a pioneer women's studies practitioner, is the author of a key analysis of undergraduate education, one that has informed this and other curriculum- and pedagogy-reform projects: *They're Not Dumb, They're Different: Stalking the Second Tier* (1990).

## NOTES

1. Here I am borrowing the language of Catherine Stimpson, founding editor of *Signs: A Journal of Women, Culture, and Society*.
2. I am thinking particularly of Elizabeth Fennema and her Wisconsin-based research group.
3. This research includes work by myself, the Sigma Xi organization, and Elaine Seymour's study of students who leave the sciences well into the major.
4. I called it a "hemorrhaging" in my book *They're Not Dumb, They're Different* (1990).
5. *Investing in Human Potential: Science and Engineering at the Crossroads* (1991).
6. Dennis Bartels and Judith Opert Sandler, eds., 1995.

# Preface

**Cheryl Ney, Jacqueline Ross, and Laura Stempel**

> Our science would not be elite and authoritarian and, therefore, it would have to be accessible—physically and intellectually—to anyone interested. It would be humble and acknowledge that each new "truth" is partial; that is, incomplete as well as culture-bound. Recognizing that different people have different experiences, cultures, and identifications (therefore, different perspectives, values, goals, and viewpoints), feminist science would aim for cultural diversity among its participants, so that through our diverse approaches we would light different facets of the realities we attempt to understand. Such diversity would help to ensure sensitivity of the scientific community to the range of consequences of its work and thus its responsibility for the goals of science and the applications and by-products of its research.
>
> —Ruth Bleier[1]

Collaboration in the scientific community is a well-established means of obtaining results that no single theorist or researcher could produce. This book explores the value of collaboration when applied to the field of education: collaboration among students in the classroom, among faculty colleagues, between faculty and administrators, and among all of those who are interested in revitalizing the way that science is taught and the way students learn. Academic disciplines can be transformed by collaborative work, and collaboration has been at the heart of large-scale curricular and pedagogical reform movements as exemplified by Women's Studies. In the sciences, collaboration is the usual method by which practitioners—working scientists—proceed. The premise of this book is that it should also be at the heart of the methods by which the sciences are experienced in the classroom.

Because teaching is the central mission of higher education, it is crucial for everyone involved in that enterprise to value the *practice* of teaching and the scholarship related to it. This book demonstrates how faculty can include pedagogical issues in their scholarship efforts as they work to develop a more reflective teaching practice. By endorsing extensive collaborations at every level, from faculty members in the classroom and researchers in the laboratory, to engaging in scholarship within their own departments, to collaborations between departments on an individual campus and between institutions, this book encourages educators to reach out. The results of these collaborations affect the ways in which science is regarded and taught, which in turn leads to systemic changes in content, pedagogy, and climate. The hope is that those systemic changes will increase the participation of underrepresented groups and also improve the learning opportunities for majority students.

While a feminist critique of conventional science and science learning informs this book, *Flickering Clusters* does not itself consider that critique in detail, nor does it provide an analysis of the national science curriculum or a study of the status of women in the sciences. (The literature review in Appendix B contains references for those interested in further study of either of these subjects.) This book is not a simple "how-to" guide to changing the way that science is taught at the college level. Instead, the book uses the *discovery method* to lead readers to an understanding of how the strategies applied herein can be taken back to their classrooms, departments, and institutions. Briefly, as Cheryl Ney notes in Chapter 2, the discovery method is a mode of learning in which the instructor "guides the students' conceptual understanding in order to make sense of observations they have made." In contrast to the traditional "sage on the stage" model, in which the teacher imparts information directly to students, instructors using the discovery method encourage students, through observation and inquiry, to form their own conclusions. This book not only describes the ways in which this method was adopted throughout the University of Wisconsin project, but actually employs it in its structure and content. By guiding readers through their own process of discovery, rather than presenting a simple "cookbook" on how to reform classroom practice, *Flickering Clusters* exemplifies the discovery method it endorses.

We have also tried to outline the major issues and some of the potential pitfalls facing those who attempt this kind of reform. By offering concrete examples of innovations in pedagogy and course content, along with attempts to improve the climate facing women and other underrepresented students and faculty in the sciences, and by presenting this information within the context of a specific project, we hope to suggest some of the ways that such transformation might occur.

*Flickering Clusters* describes lessons learned from a curriculum reform and faculty development project undertaken by the University of Wisconsin System Women's Studies Consortium and supported by a National Science Foundation grant. (The program has now been permanently instituted in the UW System.)[2] The program's goal has been to transform science, engineering, and mathematics curricula in order to attract and retain a greater variety of students. Targeted students include members of underrepresented groups (primarily women and minorities), but also a wide range of would-be majors: those who come to college intending to focus on the sciences but who are discouraged by the culture, curriculum, or classroom atmosphere from pursuing those plans. Among this group are those that Sheila Tobias has described as the "second tier," talented and science-capable students who have for a variety of reasons decided not to pursue a major in science or math.[3]

The Women and Science Project was originally structured around semester-long (or longer) visits by Distinguished Visiting Professors (DVPs) to campuses throughout the University of Wisconsin System. Individual campuses (and, in one case, a community of three campuses) developed proposals to bring in DVPs under a National Science Foundation grant. Once those proposals were accepted by project administrators, several faculty members at each site were appointed as Faculty Fellows. The DVPs taught courses in their academic specialties, organized workshops for local faculty and staff, and as a central component of their visits, worked to develop or to expand existing communities among the faculty, staff, and administrators in the sciences, math, and engineering. They served as role models in their teaching by inviting local faculty to observe their classes, and their presence on campus provided an opportunity for colleagues within and outside of their individual departments to focus their attention on pedagogical and curricular issues they might not otherwise have had a chance to pursue. Faculty Fellows also experimented with new pedagogical approaches, presented workshops with the DVPs, and were trained to carry on faculty development and curriculum reform activities after the end of the DVP's tenure. In addition to their on-campus experiences, DVPs and Faculty Fellows traveled to other sites around the state to present their work, and took part in project-wide conferences and retreats with faculty and administrators from other campuses.

Although this particular project took place in a large, statewide university system, its lessons can be applied at institutions of every sort, from small colleges to larger universities to systems like the University of Wisconsin. Since the campuses that comprise the UW System vary widely in size and student population, faculty members and administrators at a variety of institutions will find familiar the experiences that participants describe in the following chapters. Similarly, the work of coordinating this kind of reform project throughout a large state system has national implications for the development of a network of practitioners and programs committed to women and science.

Although the experiences described in *Flickering Clusters* may be of most obvious relevance to academics pursuing similar transformations in college-level science curricula and pedagogy, we believe that this book has lessons for many other readers as well. Those interested in broad questions of curriculum reform and faculty development will find many places where the specific example of the Women and Science Project can be generalized to other fields, while readers interested in pedagogical issues should be able to apply the innovations illustrated here to their own teaching, regardless of the discipline. Scholars and teachers in Women's Studies may be particularly interested in ways to encourage their colleagues in the sciences to incorporate feminist pedagogical strategies into their courses.

*Flickering Clusters* itself is a collaborative undertaking in every sense of the word. The book has been written and edited as a collective task among the three co-editors. While each of us was responsible for the basic writing of individual chapters, as a group we debated the structure and content for each section and for the book as a whole. Through lengthy conversations, we consulted constantly about how to tackle specific problems of writing, organization, analysis, and interpretation. We read, critiqued, and offered suggestions for revision of each others' chapters, and particularly in Chapters 3 through 6, even supplied additional text, which appears in some places with, and in others without, direct attribution.

This kind of collective writing is not unprecedented, especially among feminist scholars and activists. For instance, the authors of the groundbreaking work *Our Bodies, Ourselves* used this method, and many other theorists and critics have taken a similar approach in composing introductory essays or transitional material for jointly edited books.[4] What is unusual about *Flickering Clusters*, however, is that the collaboration features people beyond the co-editors and other authors named in the table of contents. It also includes the participants in the Women and Science Project, whose narratives of their teaching and learning experiences provide the content of the book.

Originally, project participants were invited to submit essays, sample classroom handouts, and other materials for what was then conceived of as a conventional anthology. Their submissions were varied in style, format, and content, and as the editors began to think concretely about how to assemble the book, it became clear that a traditional collection of individual narratives could not effectively communicate either the spirit of the project or the lessons participants learned. Instead, we decided to weave together the various voices of those involved in the program, excerpting passages from the submissions already received, soliciting new ones so that a wider range of perspectives and experiences would be included, and inviting comments and suggestions from newcomers to the program. At the end of this preface you will find a list of contributors' original essays, from which excerpts have been used in Chapters 3 through 6. Whenever those excerpts appear, they are clearly identified as coming from a particular project participant; all other narrative and transitional material has been written by the co-editors, who are also responsible for choosing which materials to include.

We took this collective approach so that the book would mirror as closely as possible the collaborative nature of the Women and Science Project itself. This unique project was carried out jointly by dozens of faculty members (primarily from science and math departments), administrators, and staff members across the University of Wisconsin System, scholar/teachers from other institutions, and evaluators from outside the project. Throughout the project, participants structured their work around two sets of parallels: how scientists work and how students learn, and faculty development pursuits and classroom activities. Similar parallels surfaced throughout the project when, for instance, Faculty Fellows found that their learning experiences in project workshops were nearly identical to those they asked students to try in "discovery"-based classroom activities.

The design of this book is meant to invoke those parallels at yet another level by weaving together participants' voices into chapters that consider questions about the Program's history, structure, content, and future. In attempting this approach, we benefited not only from the work of those who had already offered to contribute to the book, but from the faculty and staff who generously agreed to write new material or to update previously completed essays. We see the book's readers as completing another stage of collaboration as they apply the lessons contained in *Flickering Clusters* to their own lives.

Chapter 1 outlines the history and origins of the program, while Chapter 2 presents its theoretical foundations with a constructivist approach to teaching and faculty development. In Chapter 3 we illustrate the work of project participants by presenting some of the pedagogical strategies they used to revitalize their own and their students' approaches to the science classroom. Chapter 4 examines the climate in which such learning takes place and suggests ways in which specific course content can be altered or adapted to meet the needs of both underrepresented and more traditional science students. Chapter 5 considers faculty collaboration in more detail, exploring the kinds of collaborations required for a project of this sort, and by focusing on the three-campus Collaborative Community in order to supply some concrete examples, demonstrates how the collaborations were established and fostered. Chapter 6 looks at the future of the program and its initiatives by describing both how change has been institutionalized within the UW System and what work remains to be done in meeting the goal of inclusive science teaching. The appendices provide additional supporting material: in Appendix A, a report by outside evaluators on the program's effectiveness; and in Appendix B, a literature review and bibliography.

Neither this book nor the program from which it stems could have been developed without the members of our Women and Science community; some are represented in this volume and some are not, but all deserve our thanks and respect for their contributions. To the chancellors, provosts, and other administrators on each UW System campus; to David J. Ward, UW System Senior Vice President for Academic Affairs at the time this book was written; and to Stephen R. Portch, Chancellor of the University of Georgia System (and UW System Senior Vice President during the initial stages of the project), we owe a special debt of gratitude for their support.

Rebecca Armstrong, former Director of the Women and Science Program, and Nancy Mortell, program assistant, deserve particular mention for their valuable and dedicated service to the project. For taking the time to read this manuscript and for their valuable suggestions, we would like to thank Daina McGary, Dean of the College of Arts and Sciences, Capital University; Fran Garb, Professor of Biology and Academic Planner, UW System; and Frances M. Kavenik, Professor of English, UW-Parkside. Phyllis Holman Weisbard, UW System Women's Studies Librarian, and Earl Peace, Association Director of a major curricular reform project in UW-Madison's Department of Chemistry, merit special commendation for their unfailing ability and willingness to serve, each in a singular way, as important resources to the project. We also wish to thank Donna Chan Fisher, Laura Hansen,

Pat Klug, and Cate Irsfeld for their invaluable skills as we worked to put this manuscript together. In addition, we would like to thank Jo Futrell for her efforts in formatting this book for publication. Finally, we would like to express our gratitude to the National Science Foundation for its support and, in particular, to Karolyn Eisenstein, Barbara Brownstein, and Hal Richtol for their continuing advice and encouragement throughout the course of the project.

Cheryl Ney
Jacqueline Ross
Laura Stempel

## Contributors' Essays

In addition to the titled sections written for Chapters 3, 4, and 6, we have drawn upon the following essays by project participants:

Andrew Balas, "Experiment in Diversified Instruction"

Brian T. Bansenauer, "Teaching Summary and Experiences Spring 1994," "A Math Journal"

Danielle R. Bernstein, "Special Challenges in Computing: My Experience as a DVP in Computing Information Systems at UW-Stevens Point"

Heidi Fencl, "Teaching Introductory Physics for Biology Majors: A Student-Centered Approach"

Janice Gehrke, "Evaluation vs. Assessment"

Marc Goulet, "Teaching in the Wake of the Women and Science Program"

Judith E. Heady, "Experiences as a Distinguished Visiting Professor in the Biology/Microbiology Department at the University of Wisconsin-LaCrosse"

Sheue L. Keenan, "Women and Science Program: Collaborative Community"

Vera Kolb, "Journaling as Applied to Vicarious Teaching"

Sandra K. Madison, "Computer Journals"

Sandra K. Madison and James Gifford, "Fostering Inclusion Using Concrete Models"

Sharon Nero, "Community Development"

Barbara Nielsen, "Using Student Journals to Promote Learning in an Analytical Chemistry Course," "Using Role-Playing Activities in a Science Classroom"

Sherrie J. Nicol, "My Role as a Distinguished Visiting Professor," "Initial Development Plans," "Using a Mathematics Journal"

Neal H. Prochnow, "Reflections on the'Collaborative Community' at UW-River Falls, UW-Eau Claire, and UW-Stout"

Charles W. Schelin, "Promoting and Perpetuating the Goals of the Women and Science Program"

Rhonda J. Scott-Ennis, "Trainer of Trainers Development Process"

Ethel Sloane, "'Circuit Rider' to the University of Wisconsin Centers"

Alex Smith, "The Discovery Approach to Teaching College Algebra"

Loretta J. Robb Thielman, "Experience of a Faculty Fellow in the Collaborative Community," "Refocusing from Teaching to Learning"

## Program Participants

### Distinguished Visiting Professors (home affiliations in parentheses)

Danielle Bernstein, UW–Stevens Point (Computer Science Department, Kean College, Union, NJ)

Judith E. Heady, UW–La Crosse (Department of Natural Sciences, University of Michigan-Dearborn)

Vera Kolb, UW–Madison (School of Science and Technology, UW–Parkside, Kenosha)

Cheryl Ney, UW–River Falls (Chemistry Department, Capital University, Columbus OH)

Sherrie J. Nicol, UW–River Falls (Department of Mathematics, UW–Platteville)

Sue Rosser (Center for Women's Studies and Gender Research, University of Florida)

Ethel Sloane, UW Colleges (College of Letters and Science, UW–Milwaukee)

### Campus Coordinators

Gwendolyn Applebaugh, Department of Mathematics, UW–Eau Claire

Sheue L. Keenan, Department of Chemistry, UW–River Falls

Rachel Mahun (Co-coordinator), Women and Engineering Program, UW–Platteville

Robert P. Morris, Department of Mathematics and Computing, UW–Stevens Point

Sharon Nero, Social Science Department, UW–Stout

Paul Treichel, Chemistry Department, UW–Madison

Laraine M Unbehaun, Biology/Microbiology, UW–La Crosse

## Faculty Fellows

*UW–Eau Clair*
Michelle Kettler, Biology
Alex Smith, Mathematics
Andrew Balas (unofficial), Mathematics
Brian T. Bansenauer (unofficial), Mathematics
Marc Goulet (unofficial), Mathematics

*UW–La Crosse*
T.A.K. Pillai, Physics
Jerry D. Davis, Biology/Microbiology
Rudy G. Koch, Biology/Microbiology

*UW–Madison*
John W. Moore, Chemistry
Cathy Middlecamp (Academic Staff Fellow), Chemistry

*UW–Platteville*
John Krogman, General Engineering
James Hamilton, Chemistry

*UW–River Falls*
Pamela A. Katzman, Mathematics and Computer Systems
Kim Mogen, Biology
Barbara Scheetz Nielsen, Chemistry
Rhonda Scott-Ennis (unofficial), Chemistry

*UW–Stevens Point*
Gerald E. Gau, Mathematics and Computing
James Gifford, Mathematics and Computing
Sandra Madison, Mathematics and Computing
William Wresch, Mathematics and Computing

*UW–Stout*
Janice Gehrke, Biology
Loretta Jean Robb Thielman, Mathematics

*UW Colleges*
Ned Grossnickle, Biology, UW–Marathon County
Elizabeth Hayes, Biology, UW–Fond du Lac
Carla Keller, Biology, UW–Waukesha

## NOTES

1. Ruth Bleier, "Introduction," *Feminist Approaches to Science* (New York: Pergamon Press, 1986), p. 16.

2. The University of Wisconsin System includes fifteen institutions, from doctoral universities to comprehensive and two-year campuses, plus Extension. The Women's Studies Consortium is the organization of the women's studies programs at all of these institutions.

3. Sheila Tobias, *They're Not Dumb, They're Different: Stalking the Second Tier* (Tucson: Research Corporation, 1990).

4. For example, The Personal Narratives Group, co-editors of *Interpreting Women's Lives: Feminist Theory and Personal Narratives* (Bloomington: Indiana University Press, 1989).

*Chapter 1*

# Flickering Clusters

**Jacqueline Ross**

> One . . . objective . . . is to understand the methods of scientific investigation and to dispel the myths about what science is and how scientists "do science." If we can demystify science and medicine and explore the ways in which scientific knowledge is gained, we may be able to instill in our women students confidence in their own abilities. This can break down some of the barriers, real and perceived, that lead women to think that they are better off in language and literature than in science and math.
>
> —Ethel Sloane

This is a book about flickering clusters. To chemists, "flickering clusters" is a model that explains how water molecules behave in a liquid state: they come together and interact momentarily in clusters. The molecules then leave the clusters and move on to others. Flickering clusters is a useful model for explaining surface tension, capillary action, a high boiling point, and other properties of water in a liquid state. Adopted as a metaphor, the term aptly defines the "flickering" collaborations that have characterized the genesis, development, and aspirations of the University of Wisconsin's Women and Science Program. Intrinsic to the program have been collaborations at several levels—between faculty, staff, and administrators within and among institutions. In the course of the project, these collaborations extended their range to include an evaluation team, system administrators, National Science Foundation representatives, and, eventually, a nationwide network of Women and Science colleagues. The purpose of this book is to illustrate, by recounting and analyzing our collaborative experiences, how we believe positive changes can be effected in the ways in which science is regarded and taught in the academy.

The Women and Science Program was the result of the collaborative efforts of charter members of the UW System Women's Studies Consortium, which serves as the formal organization of the Women's Studies programs in all of the fifteen degree-granting institutions in the UW System. In 1989, the Consortium identified curricular reform as one of its major goals and, because of the challenges presented by the sciences, designated women and science as a focus area within that goal. The intent of the Consortium's Executive Committee was to create a model program for curricular transformation of the sciences in the UW System that could be replicated or adapted by other universities and colleges, large and small, public and private, nationwide.

As director of the Consortium from 1989 to 1999, I have been involved from the outset in the development of the Women and Science Program. We considered the major purpose of the program to promote systemic change in the way science and science education was regarded and carried out in the UW System. Within this purpose, our goal was to attract and retain women and minority students in science, mathematics, and engineering by improving the quality of undergraduate education for both women and men. In so doing, the program has sought to reverse the attrition from science among women and minority students at a point at which it is most acute in higher education: the introductory courses in the undergraduate science curriculum. Additionally, we intended to forge links between the sciences and the campus-based Women's Studies programs through interdisciplinary and team-taught courses and other extracurricular activities. As it turned out, we were, to put it mildly, naive in our aspirations, particularly in light of our intention to complete the process within a couple of years. If we had been more wary, however, we would have striven for and achieved less.

The original Women and Science Project, funded by a major multi-year grant from the Undergraduate Education Division of the National Science Foundation, was formally launched in the fall of 1992. Faculty from across the UW System came together to discuss improving the quality of education in the sciences, defined broadly to include mathematics, engineering, and technical fields of study. Over the next few years, Distinguished Visiting Professors (DVPs), both internal and external to the UW System, in various science fields visited Women and Science communities on a number of UW System campuses. Hundreds of science faculty from all UW institutions participated in workshops, conferences, retreats, and other development activities designed to demonstrate and discuss new gender-sensitive strategies for improving the curriculum, pedagogical approaches, and climate in the science classroom. The operating principle was to apply what had proved successful in Women's Studies courses to those in the sciences.

## Background: The Legacy of Ruth Bleier

As I have already indicated, much of what we've learned in our Women and Science Program could be applied to efforts in other colleges and universities. That our efforts began when they did can be attributed to the strong tradition of both Women's Studies and women and science in the UW System. Epitomizing this dual tradition and the links between them was Ruth Bleier, Professor of Neurophysiology in the Medical School and of Women's Studies at UW-Madison until her death in 1988. In addition to being a founding mother of that program, she was a Doctor of Medicine.

But her accomplishments do not end there. Ruth was also a feminist activist, committed to promoting positive change throughout the university system and beyond. She was, in her own words, "an agitator of the university administration and an organizer of women at the University of Wisconsin beginning in 1970,"[1] who "worked consistently toward the establishment of women's studies on campus and the UW System, a goal that was accomplished in 1975."[2] It was in this role that I came to know Ruth Bleier in the 1970s. As a young faculty member in one of the comprehensive institutions in the UW System, I was mainly intent on developing my teaching and research with tenure as my goal. While I had already been affected by discrimination as both a student and a faculty member, I was blissfully unaware of or had shut my eyes to such experiences.[3] It was at this time that Ruth was visiting campuses across the state, raising the consciousnesses of women faculty, staff, and students in every institution in the UW System with multiple results. I was, perhaps, typical among my women peers (few that they were) in that I began to become aware of the promise of Women's Studies and feminist scholarship, both in terms of my own professional goals and of working with other women in the UW System to achieve common goals. Joining with Ruth in her organizing efforts, UW women formed a systemwide Coordinating Council of Women in Higher Education and, by 1975, a network of individual Women's Studies programs in each of the fifteen institutions in the UW System. That network has continued to function and, in 1989, became officially constituted as the UW System Women's Studies Consortium.

As the author of three definitive books on the hypothalamus in animals, the germinal work *Science and Gender: A Critique of Biology and Its Theories on Women*,[4] and other writings, Ruth Bleier's contributions as a scientist and, in particular, in the area of women and science have been widely known for many years. At UW-Madison, she was instrumental in the formation of the October 29th Group, a circle of scientists and other women which met over a period of years "to discuss and define a feminist critique of science."[5] In their introduction to her last address, delivered for her on January 4, 1988, a month before she was to die of cancer, Judith Walzer Leavitt and Linda Gordon paid tribute to Ruth's contributions as a beloved colleague and feminist activist and, on a national and international level, as a scholar:

> Among the first scholars in the United States to examine critically the foundations of the modern biological sciences from a feminist perspective, Ruth has provided important insights and direction to other scholars. Women's Studies courses around the country use her articles and books to provide students with core understandings of the complex issues regarding women's nature, biological determinism, and the nature of sex differences. Based on her own scientific work in neurophysiology as it relates to the biological sciences, as well as to psychology, sociology, political theory, and anthropology, Ruth Bleier's work was truly interdisciplinary and integrative. (p. 183)

Unfortunately, Ruth's impact on the mainstream of science as practiced in the academy was not as dramatic as upon Women's Studies. I recall few scientists in the packed lectures I attended, where she talked about the implications of gender and science, but Ruth indicated that these audiences were typical in her experience. One possible cause of the lack of interest among scientists was the paucity of women in those fields on the campuses. The major reason, I believed, was the suspicion engendered by her research in gender and science. The suggestion that there might be an affinity between Women's Studies and the objective sciences was either not to be taken seriously or viewed with hostility. Ruth's analysis of the problem is more complex and continues to be relevant today.

In her last address, describing the reasons for the resistance of science to feminist perspectives, Ruth observed that the latter posed major "challenges to positivism, a bedrock principle of Western epistemology, and to the objectivity and value neutrality that make of science, in our society, the best if not the only route to knowledge." Moreover, "feminist analyses . . . have implications for the gendered identity, structure, and content of science, as well as implications for science's role in legitimating society's most cherished gendered beliefs and structures, namely, that hierarchical gendered social structures are based in differently gendered human natures." Finally, Ruth emphasized her belief that "there is a relationship between 'scientific' theories of gender differences and of the inferiority of women and the virtual exclusion of women as colleagues and equals from most science department and other science institutions" (p. 193). As a result of these beliefs, the vast majority of scientists, including most of the small pool of women scientists, have never been presented with a feminist critique of science—at least, without negative connotations. I have been told by several of these women that, to this day, an expression of interest in the critique would jeopardize their careers.

Interestingly, Ruth told me when we first met that, while dedicated to Women's Studies and women's issues, she had initially been skeptical about their application to science. Gradually, however, she became convinced that gender issues were, in fact, significant concerns for scientists and science. In an essay published in 1986, she began by asking: "What is it about science—or about women—or about feminists—that explains the virtual absence of a feminist voice in the natural sciences, as an integral part of the sciences, with the single exception of primatology? And what would such a voice sound like? How would science be different? How would our perceptions of the natural world, of women and men, be transformed?" She went on to observe ruefully: "While over the past 10 to 12 years, feminists within science and without have been dissenting from and criticizing the many damaging and self-defeating features of science (the absolutism, authoritarianism, determinist thinking, cause-effect simplifications, androcentrism ethnocentrism, pretensions to objectivity and neutrality), the elephant has not even flicked its trunk or noticeably glanced in our direction, let alone rolled over and given up" ("Introduction," p. 1).

Not to be deterred, however, Ruth continued to speak out, write, and pursue in many ways her goals as a feminist and scientist until her death. From 1982 to 1986, she served as Chair of UW-Madison's Women's Studies Program. As Sue Rosser describes in a tribute to Ruth Bleier, she was

> unique among scientists who are feminists in that she did not leave her feminism at the laboratory door. She used her feminist analysis to critique existing theories of science, to point out racist, sexist flaws in experimental design and interpretation to begin to sketch the parameters for a feminist science. Perhaps more importantly, she brought the feminist critique to bear on her own research and that of her colleagues in neuroanatomy. . . . Ruth was unusual among feminists in that she continued to be a practicing scientist while working on and writing about feminism and science during the four years that she chaired the Women's Studies Program.[6]

How, then, did Ruth define a feminist approach to science? In doing so, she outlined some of the underlying principles, while noting that the list was in a formative stage and hence inadequate. Highlighting that list was the following:

> We would first of all insist that scientists acknowledge that they, like everyone else, have values and beliefs, and that these will affect how they practice their science. The next task is to convince them to explore and understand in which ways these subjectivities specifically affect their approaches, their actual scientific methods. . . . Our science would not be elite and authoritarian and, therefore, would have to be accessible—physically and intellectually—to anyone interested. . . . For many of these changes to occur, scientists would have to learn to reconceptualize science, its methods, theories, and goals, without the language and metaphors of control and domination. ("Introduction" pp. 15-16)

She also observed that "one might say that what I have described is simply good science, not just feminist science. That is true, in the same way that feminist scholarship in all fields has made them better, opened them to new perspectives and to previously ignored experience, and radically introduced gender as an unavoidable category of analysis" ("Introduction," p.17).

She concluded this essay with the challenge: "Whether feminism can bring other profound transformations in what we call science is a question to be answered over the next few years" (p. 17). Her final address concluded with this more optimistic statement: "the work has begun and progress will continue to be made as more women scientists and feminist scholars, no longer awed by science, become engaged in the task" (p. 195). With Ruth's untimely death in 1988, one month after these words were written, the task of carrying out and completing the work she had begun, of refining and building upon these principles, was left to others.

While Ruth Bleier's challenge was clearly directed to a larger sphere than the University of Wisconsin, the implications of her legacy have continued to reverberate here, in particular. Related are the challenges presented by the application of gender-sensitive approaches to teaching science, which, as the succeeding chapters of this book demonstrate, have been formidable. I have focused on Ruth Bleier to pay tribute to her achievements, for surely our Women and Science Program is rooted in and thus indebted to her work and thought. This focus also articulates the ideas underlying the genesis, rationale, and development of the program. For example, the program is grounded in the belief that the status of women in science is integrally connected to stereotyped views of gender differences, in a political context. For this reason, we focused on using female-friendly pedagogies and other Women's Studies techniques in working toward increasing the representation of women in the sciences.

Revisiting her work also reminds us of her stunning accomplishments and the paths she set forth that remain unexplored. Unfortunately, many of the issues and concerns she raises are true of the way science is regarded and practiced in the academy in the decade since her death in 1988. Moreover, the important relevance of gender to science is still understood by a relatively small number of feminist scientists and scholars. With so few Women's Studies practitioners in science, the gap between Women's Studies and science disciplines remains. And although the numbers of women who choose to become and remain as scientists or teachers of sciences have increased, the challenges remain.

## Background: Development of the Program

It was within the context of Ruth Bleier's legacy that the first priority of the newly formed UW System Women's Studies Consortium was the development of a women and science program to address the paucity of women in the sciences. Relatedly, we proposed to address what was then identified as a "pipeline" problem—that is, what prognosticators, in their wisdom, saw as a decline in the number of scientists who would be available to meet the future needs of society. Since the majority of new entrants into the workplace in the early 21st century are women and minorities, it seemed to make sense as national policy to encourage members of these underrepresented groups to enter science fields as a profession. A number of reports by prestigious national commissions and science organizations issued strong recommendations in this regard. Thus, funding agencies began to make grants available to meet the goal. It is by now commonly acknowledged that the pipeline theory, for various reasons, did not hold true. Rather, scientists, not unlike academics in other fields, have begun to face shortages in external funding and, relatedly, in teaching and research positions.

However, we in the Consortium considered the window of opportunity presented by the presumed pipeline problem, designed a curricular development model project to address it, and sent it in to one of the federal funding agencies. It is important to note that, then as now, we saw the problems as more complex and profound than those concerned mainly with workforce issues. Having been steeped in the ideas of Ruth Bleier and other feminist scholars, we were concerned by the apparent lack of opportunity for women in the sciences in academia and the world of work. Reflecting the trends at the national level, there were very few tenured women faculty in science fields throughout the UW System. Moreover, we were concerned about the stories we had heard about the hostile climate in the sciences affecting women faculty, staff, and students. These stories seemed to confirm the continuing resistance in the sciences pointed out years earlier by Ruth Bleier.

In light of this resistance, it is not surprising that we encountered barriers from many faculty in the sciences as we developed the program in their territory. Not all of the resistance came from scientists, however. The first time that we submitted our proposal based on our women and science model to a well known agency in higher education, it received excellent marks from the reviewers and high praise from the agency staff contact but was rejected by the director, whose comment I still remember: "Science is like pornography," he said; "I know it when I see it, and this is not science." We were interested to learn that the director was a humanist, not a scientist. In contrast, we received a very positive response from the director of the National Science Foundation program which eventually funded our project—a scientist whose views had been informed by a workshop on curricular transformation (incorporating material on race and gender) he had attended at the University of Maryland.

Like many other institutions, we had already brought women scientists to our campuses to give lectures and had carried out successful Women and Science Days for middle school and high school girls. Unfortunately, while such isolated activities raised awareness to some degree on gender-related issues in the sciences, they had a very limited impact on the institutions and virtually none on the teaching in these fields. Our task was to develop a project where change in pedagogy, content, and climate could be instituted and sustained in science departments within institutions and throughout the UW System. We did not realize how formidable a task this was. While it might have been better if the project had originated with science faculty and administrations, such a prospect was unlikely at this point in our history. Few, if any, departments had the critical mass of enthusiastic and committed tenured faculty necessary to initiate and carry out such change. And, while it would certainly have been helpful if we'd had more representation from the sciences among our women's studies faculty, this was not the case. However, we moved ahead anyway.

The project was a collaborative one from the outset. In the design as well as through the course of the project, we consulted with and examined the ideas of a number of teacher/scholars concerned with gender issues in sciences. For example, early in the process, the feminist biologist Anne Fausto-Sterling gave us a valuable critique of our project grant proposal. We also were assisted by reading works by Sue Rosser and Sheila Tobias and discussing with them ideas for applying feminist approaches in the classroom. Sandra Harding's perspectives on feminism and science in her writings and in a talk at one of our annual Women's Studies conferences, which focused on the subject, generated a great deal of discussion and controversy.[7] We also talked with faculty around the country who were trying out active learning, inquiry-based, and other methods in the classroom. Additionally, we exchanged ideas with other women and science programs. Many of them, we learned, focused on recruitment and retention of women with little if any attention to approaches grounded in feminist critiques of sciences and applied to curriculum and faculty development. They had been most successful, it seemed, at addressing problems in these areas through mentoring and other activities. However, those involved with such programs were a valuable resource to us in planning and carrying out our program. Finally, we talked to faculty and administrators in our own institutions. From many of these individuals and organizations, we gleaned insights and ideas that were adapted to and woven into our program.

While there were a number of commonalities among their approaches to recruiting and retaining women, rarely were there links between the sciences and Women's Studies. Indeed, there was often suspicion and hostility. Fortunately, there were a few strong women feminists in the sciences, but they were definitely in the minority. From all accounts, they tended to feel a great deal of negative pressure because of their views, both from their female and male peers. Not only did many if not most of the scientists find the notion of a feminist critique of science repugnant, they were disdainful of what they feared might be "soft" or "touchy-feely" techniques in the classroom. Yet, we were, in Ruth Bleier's terms, promoting "good science."[8] For their part, many women scientists themselves seemed to shy away from the "f" word or any ideas that might be construed as radical or "male-bashing." A major challenge was to convince them that we were not proposing watered-down science.

A second major challenge was to bring to their attention the value of the feminist critique of science. As Distinguished Visiting Professor Cheryl Ney explains it, for many if not most women scientists, the problem involves lack of awareness of how science shapes ideas of femaleness. Most don't understand that women have been excluded from research studies on health prior to the 1970s. Furthermore, they do not recognize how feminist perspectives can help to develop explanations for natural phenomena. Many women scientists are unaware of feminist scholarship and its relevance to their work.

She contends, "a critique is intended to enhance or improve upon that which is being critiqued. And so it is with the feminist critique of science. What the feminist critique chiefly does for our understanding of science is to remind us that human beings do science—human beings who are historically located and culturally socialized with unique life experiences. And, in science, these human beings have developed methods to observe the natural world and test out their understandings of this world. If it is accepted that the aim of science is to develop explanations that work to explain our observations of the natural world, then we can see where our humanness influences our science. Explanations are a product of human minds—minds influenced by experience."[9]

Thus it is, as Ruth Bleier has pointed out, that gender can have a significant influence on the practice of science. In her last paper, for example, she discussed how she and other feminist scientists had questioned the assumption that science is objective and value-free. Their work, in fact, reveals "the gendered nature of the body of knowledge we call science . . . ." As a prime example, she cites the scientific efforts which purport to "demonstrate that sex differences in the structure and function of the brain underlie presumed gender differences in cognitive abilities" (p.190). Her own work, based on a study of the human corpus callosum, refuted these efforts, pointing out their conceptual and methodological flaws.

Ney goes on to say,

> The feminist critique suggests that women, who do indeed have minds, can bring their experiences to bear on explanations in science. Different perspectives may give rise to different explanations, but all of these explanations are subject to the scrutiny of the standards for acceptable explanations in science. By including wider perspectives in Science . . . a wider range of male and female perspectives, more explanations will be considered, and the sciences will be enriched because of this.
>
> In addition to providing a wider diversity of perspective in the shaping of scientific understandings, women have made other unique contributions to science. It is well established now that women, like men, choose to investigate and spend their lives in research about those things of interest to them. For example, it is no wonder that as more women have gone into the field of medicine, we now have a field of medicine that is Women's Health. Including people of differing experiences in science widens the realm of what is investigated in science—and how investigations take place.
>
> It is also widely accepted now that women have made unique contributions to the methodology of various disciplines in science. The careful examination of Barbara McClintock's life and work by Evelyn Fox Keller has revealed McClintock's unique method of having "a feeling for the organism"—and this led to her revolutionary understanding of gene transposition. Other examples of unique methods interdicted by women have been documented. Martha McClintock, a biopsychologist, was an early proponent of studying rat behavior with not isolated rats (the prevailing method) but rather with rats in the presence of other rats![10]

In discussing the impact of female primatologists in refuting prevailing assumptions about male dominance in the organization of primates, Ruth Bleier emphasized that in this field "with a high proportion of women researchers who raised new questions from a self-consciously female and feminist viewpoint, dominant paradigms were toppled and fresh research directions were taken" (p. 187). As a result, the knowledge in this field has become richer. This is but another example of the importance of questioning the validity of prevailing interpretations of nature through a feminist critique.

The original UW Women and Science project was developed based on this foundational understanding that comes out of the feminist critique of science. A wider acceptance of a diversity of perspectives in science will lead to an enriched science—with enhanced understandings, a wider scope of investigations and approaches to research. Developing this wider diversity of perspectives requires an examination and a revitalization of the science curriculum for undergraduates, beginning with introductory science courses. Additionally, the development of the project was influenced by Sheila Tobias' concept of the "second tier," those students who are capable of doing science but choose not to. It made sense to us that both female and male students would benefit from the new perspectives and insights derived from our application of the feminist critique. Furthermore, Tobias' theory was useful in persuading faculty and administrators, who might have been suspicious of a women and science program per se, to give us a chance to pilot our project.

As indicated above, our project was designed to reform those aspects of introductory courses—specifically content, climate, and pedagogy—that Tobias had demonstrated discourage capable men and women students from further study in the sciences. The program intent was to reform introductory curricula and increase female and minority representation in science by: a) increasing faculty expertise in gender and science scholarship and pedagogy; b) creating role models of professional women scientists; c) improving classroom and campus climate; and d) creating "science communities" that would promote effective learning. Since such innovations had been shown to be attractive to white men as well as to women and people of color, the project, we argued, should gradually effect improvements that would increase the total number of students majoring in science. Finally, the purpose of the project was to promote permanent systemic change in science education in the UW System.

It was our intention to bring students and faculty at host campuses together with Distinguished Visiting Professors (DVPs) of Women and Science who had successfully implemented teaching innovations at their home institutions. Since we had initially and naively thought our goals could be reached in three years, we had envisioned no more than six Distinguished Visiting Professors, a number that needed to be revised as we expanded the time frame of the project. By the conclusion of the project period, each of the DVPs was to visit a "science community" made up of one or more science departments in one or more UW System institutions, teach a model class, and work closely with a small groups of Faculty Fellows (FFs) as well as with larger groups of faculty and staff. The host communities were chosen through a competitive process. We sent out a call for proposals to all UW System institutions, asking for projects involving one or more departments, colleges, and institutions, that would benefit from the presence of a DVP. At the same time, we solicited nominations for two or three Faculty Fellows per site, with the stipulation that at least one of them had to be tenured.[11] Finally, we asked that the institution match our funding; the total amount supported the DVPs, FFs, and supplies and expenses related to the project.

The typical Distinguished Visiting Professor would spend a full semester at a University of Wisconsin "science community," teaching a model introductory science or mathematics course, holding seminars on the incorporation of race- and gender-sensitive content into introductory science teaching, and working closely with Faculty Fellows from the host communities to develop new course materials and syllabi. Each Fellow was required to develop a new or revised course and teach it within two years. The program also proposed to develop a cadre of faculty development consultants within the UW System, some of whom would serve as future DVPs and others who would facilitate workshops on other campuses. And, as mentioned earlier, a Women and Science Advisory Board, drawn from each of the campuses, was to help guide the direction of the project. The project was administered by a Program Director, Rebecca Armstrong; I was the principal investigator and oversaw its operation. While our roles were somewhat different, we nevertheless worked as a team, along with the campus participants, to navigate the program toward its goals.

## Launching the Program

The Women and Science project began in 1992 with a conference whose program included a plenary address, mini-workshops, and introduction to the project. Sheila Tobias, the plenary speaker, presented her views of how and why curricular reforms should target female students in science and the difficulties in achieving this kind of change. Following her talk were workshops presented by faculty from around the UW System, offering examples of successful courses and programs addressing gender and undergraduate science education. Those in attendance seemed to welcome the rare opportunity to network and share ideas with colleagues on the topics which, at the time, were rarely discussed on their campuses. Interestingly, an overwhelming majority of the participants identified warming the classroom climate and developing a sense of community as the most important elements in attracting more students—including women and minorities—to the sciences. As the program progressed, this initial judgment was to be repeatedly expressed by faculty throughout the system.

In the spring of 1993, Distinguished Visiting Professor Ethel Sloane of UW-Milwaukee, known for her pioneering work in the field of the biology of women, taught her course by that name to a group of students at the UWC-Waukesha. As she did so, three Faculty Fellows from the UW Centers, the freshman-sophomore institution in our system, now known as UW Colleges, observed her teaching strategies and worked with her to integrate such approaches into their classrooms. Additionally, she facilitated a series of mini-conferences for Centers' faculty. All this did not happen, however, without difficulty.

## Barriers/Resistance

According to the program evaluators, who conducted a program site-visit at UWC-Waukesha, one of the faculty members, initially at least, felt "threatened" by what he perceived as a "conflict with the course [in human sexuality] he was teaching." He and some other members of the department made this first semester challenging in many ways. There were little things, but they made our jobs difficult. For example, we learned just before the beginning of the semester that Ethel's course had not been announced and that no one had signed up for it. Some faculty, concerned about what they viewed as conflict or competition, were loathe to promote the course and, in subtle ways, boycotted the program. With the help of the campus administration, we developed fliers and posters advertising the class, and it was filled in time for the beginning of the term. Thus, even in its early stages, the Women and Science community formed a shield to prevent recalcitrant detractors from disrupting its order.

The learning experience in Ethel's classroom manifested the kind of feminist approach to science discussed by Ruth Bleier. Describing her course, the project evaluators wrote: "It was important to her to try to get students to recognize 'good science,' and she felt that the course she was teaching lent itself to a positive climate. She felt very well accepted by the students. For her, respect for students was an important element in providing a positive classroom climate." Student response to the course was overwhelmingly positive. One student reported that "Dr. Sloane opens the climate and is respectful of students' needs." She was described as a "positive role model who can 'empower' women" ("Interim Report," pp. 6-7). Many of the students indicated that if they had taken the class sooner, they would have made changes in their academic decisions. And some students did, in fact, change their majors. For many of them, this was the first time in their educational experience that they had understood how science could relate to their lives and that the science classroom need not be intimidating.

For most of the faculty, Ethel's example presented new ideas for both pedagogy and course content. The evaluators reported: "Because of this program, faculty generally recognized a need to change the way they taught science and to examine their own teaching style." From my perspective, I had Ethel to thank for getting the program off the ground and to an excellent start. To me, she was the pioneer DVP, experiencing all of the frustrations and helping us to address problems for those who followed her. Because she was willing to enter into uncharted territory, we had a program. Her courage, intelligence, ability and, above all, her sensitivity, were essential to the establishment and development of the program. And, in fact, she has continued to be a strong supporter of her Women and Science colleagues, always willing to counsel young women scientists around the state and, when necessary, write letters in their behalf during their tenure struggles.

Several more DVPs paid semester-long visits to UW System campuses in the succeeding semesters. Two of them, Sherrie Nicol (Mathematics) and Cheryl Ney (Chemistry), were based at the three-campus Collaborative Community described earlier. Another DVP, Sue Rosser, followed a different model, circulating among nine UW System institutions, initiating faculty development activities. She spent an average of two to three days at each of these campuses, giving workshops and talks and meeting individually with those faculty particularly interested in curricular revision. Other DVPs included Vera Kolb (Chemistry), Judith E. Heady (Biology), and Danielle Bernstein (Computer Science), who remained in residence at individual campuses. In addition to Sloane, Nicol and Kolb came from UW campuses; the other four were external to the system; all were selected through a search and screen process.

Approximately halfway through the project period, it became clear to us that the focus of the program was evolving. As the project evaluators noted in their interim report, issued in 1994, "the purpose and methodology" of the project did not change, but its emphasis had shifted from curricular to faculty development, especially "with respect to a) the acquisition of knowledge and understanding of the scholarship on gender studies, and b) training of faculty to facilitate the institutionalization of the project" ("Interim Report," p. 2). What we came to realize was that, because systemic change would take a much longer time than we had anticipated, it made sense to concentrate, in the initial phases of the project, on building a community of faculty educated in female-friendly approaches to the sciences.

As the program progressed, we began to develop a "trainer of trainers" model which focused on preparing the Faculty Fellows to carry on the function of the DVPs at the conclusion of the project period. Thus, the Fellows, mentored by the DVPs, began to design and present their own workshops and activities, testing them out before our Women and Science community, and eventually offering them at various campuses around the state and nationally. This model evolved into a network of faculty development consultants, from which the core faculty for the 1997 Summer Institute was drawn. Our intention is to expand this network into a national Women and Science community. The annual Institute will use this network as a resource. Even more important, the network will provide its members with a vehicle for sharing information and ideas, including opportunities for further collaboration on individual and institutional bases.

## Community Building

In the course of the project, we began to concentrate on developing "flickering clusters" of interlocking communities as an extension of collaboration, as a way to address the chilly climate in the sciences and, relatedly, to promote changes in the curriculum. Building these communities with and among campuses was an organic process, beginning with a nucleus of science and Women's Studies faculty and extending to include many other faculty, staff, and students from a variety of disciplines, as well as administrators. As the numbers grew, these groups, as with flickering clusters in water, developed a strong sense of cohesiveness. To strengthen the community, we have held annual retreats for all program participants and encouraged further communications through electronic media and campus workshops. Faculty participants reacted very positively. As one stated, "There was almost no collaboration among the three colleges before this program. . . . It was affirming to see that we had common problems among the disciplines on our campuses" (quoted in "Interim Report," p. 16). As Cheryl Ney puts it, "these communities built knowledge about teaching together. The relationships established provided faculty with people they could talk with about their teaching and explain their teaching experiments as they progressed. This mirrors the research process that scientists are trained in."[12]

## Moving Toward the Future

In flickering clusters, each water molecule is "simultaneously a hydrogen bond acceptor and a hydrogen bond donor, and a sample of water is a dynamic network of H-bonded molecules."[13] Again, the metaphor is useful in describing the evolution of the role of the Faculty Fellows and other participants in the Women and Science project. Most of them, having learned from the DVPs and each other, have become "donors" or leaders in their science communities. Further, they are now agents of change sparking reform among some or many of the rest of their colleagues. Yet, it is also true that a number of participants have experienced backlash from colleagues. Some have had difficult tenure battles related to their participation in the program. Others have continued to meet with resistance to their advocacy of changes in the ways in which science is still taught by the majority of their colleagues. Still, it should also be emphasized that faculty participants, with the support of their program colleagues, have been able in many cases to stand their ground. So, for example, the Faculty Fellow initially denied tenure was emboldened to convince members of her department's executive committee to change their recommendation. She is now an associate professor and is recognized as a valued consultant on Women and Science Program issues.

And, I've been told by some former Fellows, many of the faculty who had resisted the new strategies introduced by the DVPs a few years ago are now trying them out in their classes. Through the project years, we have welcomed newcomers to the network, including the new Program Director, Heidi Fencl, a physicist. We now have a cadre of faculty development consultants drawn from the program community who are available to work individually and as teams with other institutions. Their contributions are reflected in the text of this book. In this, our next phase, we are encouraging the development of a national network of flickering clusters or science communities.

Now, in this new phase, we have begun to enhance and broaden our efforts and to reach out to colleagues across the country. In 1997 and 1998, we held our first national Women and Science Summer Institutes. We invited faculty from universities and colleges, large and small, to submit projects relevant to female-friendly pedagogy, curriculum, and climate, for development over a five-day period. The groups were comprised of both women and men drawn primarily from the sciences but also included faculty from Women's Studies and other disciplines, and a number of administrators as well. Facilitators for the institute were drawn from former DVPs and Faculty Fellows in our program. As it turned out, we were overwhelmed by the response both to our call for proposals (we could accept only a fraction of them) and to the Institute itself. From all accounts, participants were excited and pleased by the opportunity to work with fellow agents of change. What would also be useful, they informed us, is a book that could provide a framework for and help chart the development of gender-friendly science on their campuses—in short, this very book, which we were already in the process of writing.

This and other experiences reinforced our belief that we had a story to tell that could help guide our colleagues elsewhere who wished to change some or all of the ways science is presented in their institutions, ranging from the introduction of female-friendly pedagogy to feminist content to the development of women and science communities within and among institutions. Further, we have acted on the assumption that, as Sheila Tobias demonstrated in *The Second Tier*, what works well for female students will also work well with intelligent, scientifically-inclined male students who drop out of these fields because of the ways in which it is characteristically taught. In the process of telling this story, we shall demonstrate that what works with women students is beneficial for all students and is, in fact, good science.

In preparing for the next phase, we reviewed, with an eye to the future, the program's identity, its goals and accomplishments, in relation to its formal title, "Science, Diversity, and Community" and what had become its informal one, the Women and Science Program. Over the years, there had been a debate over whether or not we should abandon the latter. After much debate, the community decided that, while it was true that our strategies applied to both men and women, it was important that the program continue through its name to be identified with our original principles and goals. Thus, the project in its new phase is officially the Women and Science Program.

In the process of this review, we also had to admit that, during the project period at least, we had not by any means met our own expectations regarding the recruitment and retention of minority students and faculty in the sciences. Yet, we still hope to demonstrate in our next phase that the practices developed in the program can be effectively adapted to learning experiences in more diverse academic settings. The experiences of Catherine Middlecamp and Marc Goulet, described in Chapters 3 and 6, respectively, lead us to hope that applying such strategies affecting content, pedagogy, and climate will contribute to the development of truly multicultural sciences communities in the future.

## What We Learned

The first year of the program was, perhaps, the most difficult and instructive in ways that may be helpful to others wishing to begin programs of their own. Our design as described above was the blueprint that informed but did not constrict the program as it developed. What we learned from the outset was that while science departments at every institution shared characteristics with those on other campuses, each also had its own culture. We also discovered that it was crucial for the hosts, including faculty and administrators, to "buy into" the planning process for the project to succeed. Thus, we did not use the identical approach in building the program in each community. Rather, we tried to get all of the players from the host campus together with the DVP to plan the visit well before the beginning of the semester. The result of this kind of team effort was and should be the development and implementation of a model Women and Science Program that, while informed by and related to the larger project, is unique to each home community.

While I know that anything to do with administration (and in our case, System Administration) sets faculty on edge, we in this program have also learned that it is nonetheless an area that needs serious attention. As one of the DVPs indicated, the "[UW] System needs to see itself as a partner with the project." What this meant was that the Program Director and I worked hard to develop and maintain effective and frequent means of communication with program participants. We visited each of the host communities at least twice during a DVP's semester visit, attending many of the seminars and workshops. We tried to be responsive to the questions and concerns of the communities; needs emerged that we hadn't anticipated. For example, early in the project, it became apparent that we needed coordinators in the host communities. Working together with the campus administrators, we managed to develop arrangements to support a program participant in this role. As it turned out, the Campus Coordinators played a critical role in carrying out the program throughout the course of the project. Additionally, and this is probably no surprise, it was always an advantage when the campus administration saw itself as a partner in the project and very difficult when they did not. Administrators at all levels, from Campus Coordinator to department chair to dean and provost, should be brought into the process and urged to contribute in their own ways.

Relatedly, we were reminded repeatedly of the importance of clear communication in every aspect of the project. At the outset, we underestimated the extent to which we needed to explain and describe the goals and design of the project to faculty and administrators on host campuses. This was made clear to us by the evaluators early in the project. Although we sent out reams of materials, many of them never reached the right people or were not read. It is also true that ignorance was frequently used by those in opposition to the program to attack it, claiming that they had never heard that their campus or department was to be a host when this was simply not true. On the other hand, faculty and administrators are very busy people these days, and we soon learned that we had to find effective ways to capture their attention. By bringing them into the process of planning and implementing their program, we began to solve what had been a problem. And, by the midpoint of the project, program participants themselves were becoming creative in developing new and improved ways of communicating formally and informally.

Another very important lesson we learned early in the project is the importance of evaluation. The purpose of the evaluation is twofold: 1) to assess whether the program has met its original goals; and 2) to determine what methods have been successful in implementing it and, relatedly, if and how in the course of the project its design should be altered. Since the program was intended to serve as a national model, we decided that the evaluation should be formative, to provide insights and information not only to ourselves in the course of the project but to other institutions that might replicate some or parts of it. Because we needed evaluators with expertise in gender and science, we hired two whose combined talents have been extremely useful throughout the project—Judith Levy, then Chair of the Chemistry Department at Eastern Michigan University, and Gloria Rogers, Dean for Institutional Resources and Assessment at Rose-Hulman Institute of Technology. As they indicate in Appendix A, they carried out an in-depth assessment, which included interviews with program participants. Of utmost importance to all of us was the information we received on a regular basis from the evaluation, which enabled us to analyze, fine-tune, and, when appropriate, make changes in the program design and implementation.

We learned that change takes time and that patience is necessary. While this should be so self-evident that it is a cliche, it was still necessary to remind ourselves that we were focusing on faculty culture and the entrenched curriculum, and that dramatic transformations would not occur overnight. In fact, change would and will not occur until faculty and administrators can take ownership of the new approaches and consequent attitudes. Moreover, undertaking a project of this scope necessarily entails both strengths and weaknesses—the latter because of the magnitude of the work involved in trying to anticipate and carry out all the details of each host community while keeping our eyes on the goal of systemic change. As with flickering clusters, "what makes it difficult to predict theoretically all the properties of water is that the hydrogen bonds extend for long distances through the liquid." To carry the metaphor further, from the model of the collective behavior of flickering clusters, we can attain a deeper understanding of nature than we could from that of a single molecule. In this book, we have tried to describe and analyze our Women and Science model for insights into the nature of science in the academy.

## NOTES

1. Ruth Bleier, "A Decade of Feminist Critiques in the Natural Sciences: An Address by Ruth Bleier," with an introduction by Judith Walzer Leavitt and Linda Gordon, *Signs* 14 (1) (Autumn 1988), p. 188; subsequent quotations cited in text as "Decade."

2. Ruth Bleier, "Introduction," *Feminist Approaches to Science* (New York: Teachers College Press), p. 4; subsequent quotations cited in text as "Introduction."

3. When I first enrolled in a Ph.D. program in English at UW-Madison in the 1960s, I was asked by Kathryn Clarenbach, a prominent feminist both in Wisconsin and at the national level, whether I had ever experienced any discrimination in my field. Surprised at her question, I assured her I had not—this, despite the fact that I had never as a student at the University of Michigan, where I received two degrees in English, had a woman for a teacher. (Nor, as it turned out, did I have a woman teacher at UW-Madison.) Moreover, I had been discouraged from pursuing a Ph.D. in English from the University of Michigan by my major professor with the reason that "English is an effeminate enough subject"—hence, women should not get graduate degrees in this area.

4. Ruth Bleier, *Science and Gender: A Critique of Biology and Its Theories of Women* (New York: Pergamon Press, 1984).

5. Patricia Witt et al. "The October 29th Group: Defining a Feminist Science," *Women's Studies International Forum* 12 (3) (1989), p. 253.

6. Sue V. Rosser, "Ruth Bleier: A Passionate Vision for Feminism and Science," *Women's Studies International Forum* 12 (3) (1989), p. 250.

7. Sheila Tobias, *They're Not Dumb, They're Different: Stalking the Second Tier* (Tucson: Research Corporation, 1990), and *Revitalizing Undergraduate Education: Why Some Things Work and Most Don't* (Tucson: Research Corporation, 1992); Anne Fausto-Sterling, *Myths of Gender: Biological Theories about Men and Women* (New York: Basic Books, 1992); Sandra Harding, *Whose*

*Science. Whose Knowledge* (Ithaca, NY: Cornell University Press, 1991); Evelyn Fox Keller, *Reflections on Gender and Science* (New Haven: Yale University Press, 1985); Sue V. Rosser, *Female-Friendly Science: Applying Women's Studies Methods and Theories to Attract Students* (New York: Pergamon Press, 1990); Sue V. Rosser and Bonnie Kelley, *Educating Women for Success in Science and Mathematics* (West Columbia, SC: Wentworth Publishing, 1994).

8. I have avoided dealing with the controversy of feminism in relation to the "good science" vs. "bad science" debate (see Fausto-Sterling and Harding in the works cited above). The basis for this controversy is the assumption, on the part of some scientists, that there is no difference between what we call "feminist science" and "good science." That is, there is no recognition that feminist science is groudned in a theoretical social and political framework informed by feminism. Clearly, Ruth Bleier does recognize the importance of this framework and is saying that feminist science equals good science, although the reverse may not be true.

9. Cheryl Ney, personal communication.

10. Cheryl Ney, personal communication.

11. Because we recognized the controversial nature of our programmatic goals, we hoped that a tenured Fellow would provide the clout that might be needed to ensure that curricular reform would actually be carried out and institutionalized. Further, we thought it important that at least some of the Fellows, protected by tenure, could help support their more vulnerable junior colleagues.

12. Cheryl Ney, personal communication.

13. C. Matthews and K. E. VanHolde, *Biochemistry*. Second Edition (Benjamin Cummings, 1996), pp. 33-34.

*Chapter 2*

# Foundations of Teaching and Faculty Development

**Cheryl Ney**

My first year of teaching was a disaster. Having just completed my graduate work in DNA biochemistry in Chicago, I found myself in a general chemistry class, Chemistry 101, in central Ohio. It had been thirteen years since I had taken the course, so there was a great deal of content that I knew I was unfamiliar with. I didn't think teaching itself was going to be a problem: I had been in classrooms as a student since the age of five and besides that, I was only thirteen years older than the students I was teaching. I confidently felt that if I simply extrapolated from my own experiences of teaching and learning, things would be just fine.

It should come as no surprise that not knowing the content sufficiently, or how to teach first-generation, first-semester college students led to a less than satisfying first-year experience as a teacher!

After about three years, I had taught enough general chemistry to know what the themes and major concepts were in the course. I had mastered enough knowledge of many of those inorganic species I hadn't seen since my undergraduate years. I had even developed a few demonstrations and was using ball-and-stick models regularly in class (after I learned always to make the models and check them before going into class!). By making the assignment of asking students to reflect on their learning in journals, I had come to realize that students were living in a different world than the one I had come out of fifteen years ago.

I also became acutely aware of the differences among the students I was teaching—in abilities, in learning styles, in motivations and perspectives. Since this course was required of nursing majors, I was also made aware of the impact of gender, both mine and the students', in the classroom. I learned of the obstacles young women face in studying a subject that is culturally identified as male. Many of these students expressed a lack of confidence and fear of failure. These emotions, coupled with their inability to focus selfishly on their learning, really seemed to interfere with their success. For some, having a female instructor required some adjustments, since they had experienced male instructors in their prior science and math courses. Aided by a new awareness of just whom I was teaching, I began to make changes in the course, in an attempt to make it more relevant and meaningful to as many students as possible. But still, I was not satisfied with my teaching. I was confounded by a statement I often heard from many students, over the course of several semesters: "I really understood it when I was in class, but when I tried to do homework problems on my own, I got stuck."

In thinking about this problem, I began to examine the foundations of my own teaching. It was clear to me that I was doing a better job than before in teaching chemical concepts to a large majority of students, but not at teaching critical thinking and skills. I was not satisfied with my own skills in teaching about the discipline—that is, how to "think with" the concepts. I realized that while I was beginning to develop a teaching practice that was my own, by and large I was still "teaching as I was taught." It is this examination of foundational issues in teaching that has led me to a revitalized teaching practice, which I have found to be more effective in guiding students' learning of chemistry, including both conceptual understanding and critical thinking. In addition, it is this process of examining foundational issues that was the basis for my activities as one of the faculty development leaders in the University of Wisconsin Women and Science Project.

I began this chapter with my own personal story of developing as a teacher in order to raise some of the issues that educators in the fields of science and mathematics currently face. There has been plenty of research to show that students in the United States perform poorly in science and math, as well as to document the underrepresentation of women and minorities in these technical areas. There continues to be a call for teachers in science and math to change the way they teach as a remedy for this situation—to move from the traditional lecture approach, with labs that verify the concept presented in the prelab introduction, to a classroom approach that incorporates small-group activities, case studies, and the use of technology, with labs that are discovery- or guided-inquiry-oriented. This call is for systemic change—not simply for individual teachers here and there, but for a whole *system* of teachers. Change of this magnitude, it seems to me, requires an examination by teachers of what is at the root of teaching and learning—foundational assumptions not only about how we teach, but about how we develop as teachers.

This chapter explores the foundations of teaching science and math and their implications for revitalizing and reforming teaching so as to improve the conceptual understanding of and critical thinking in science and math for all students. These are issues that are central to the faculty development activities of the University of Wisconsin Women and Science Program. Many aspects of the original project focused on foundational issues in teaching and learning, since this is where systemic change is fostered. The chapter concludes with a discussion of another set of foundational issues, those related to the design and implementation of the faculty development activities in the teaching of science, math, and engineering.

## Foundations of Teaching

"Teachers teach as they are taught" is an old adage. It reflects the idea that teaching—or minimally, elements of one's classroom practice—is often grounded in unexamined tradition. In science, this can be seen in such classroom practices as lecture-only pedagogy, the assignment of problem sets with perhaps a recitation or discussion session for questions, and little emphasis in class on conceptual understanding and more on memorization. Assessment is often in the form of four exams and a cumulative final, with laboratory work usually being the standard fill-in-the-blank verification labs. Employing tradition as a foundation for teaching is based on several assumptions: 1) that good scientists and mathematicians are automatically effective teachers (why else would we receive so little training in teaching?); 2) that the body of knowledge in the discipline is unchanging, and therefore the curriculum and the teaching methods ought to remain unchanged; and 3) that students and teachers are static bodies living in an unchanged world. However, a foundation based on unexamined tradition does not provide for a process of change.

But if not tradition, what can teaching be rooted in? Can an understanding of the practice of science reveal anything for the practice of *teaching* science? While working on a project a scientist often refers to the research literature at various stages to complete it. Is there an equivalent of this for the project of teaching? Due to their narrow technical training, many scientists are unfamiliar with the research literature on teaching and learning in general, and on the teaching and learning of science and mathematics specifically, yet this literature can be very useful to scientists in the classroom. (The bibliography in Appendix B contains examples.) These research-based resources range from examples of how others have taught different concepts and their evaluation of the effectiveness of a particular lesson to alternative teaching methods that can be infused in a curriculum. The growing body of knowledge on student misconceptions is helpful as well. For example, knowing the average college-bound high school student's conception of atoms can give a chemistry instructor somewhere to start in a first college course in chemistry.

There are many constraints on scientists in using the research on teaching and learning in their teaching practice. Certainly time is a factor. How is one to keep up with the literature in an ever-growing scientific subdiscipline and grasp the teaching and learning literature as well? A good standard is the one suggested by Stephen Leiwan: "It is unreasonable to ask a professional to change much more than ten percent a year, but it is unprofessional to change by much less than ten percent a year."[1]

There is also a more serious limitation to using the research on teaching and learning as the grounding for teaching. This approach suggests that teaching can be improved simply by referring to the literature on teaching, and there is the implication that context doesn't matter: What works for one instructor, in one discipline, with one group of students in a particular setting ought to apply to all situations. Just like an experimental methods section in a scientific paper, an example of a teaching innovation must be adapted to one's own context. It is the process of identifying a teaching innovation in the teaching literature and making adaptations to one's own setting and teaching practice that could provide grounding for teaching. Indeed, the literature on teaching has characterized this activity as reflective teaching.

What would such a reflective teaching practice look like? In addition to utilizing the research literature, it would also use the research process. Scientists are well trained not only as researchers studying some aspect of the material world, but in the process of research. Why not turn one's research skills to the classroom and see each class session or course as the "experiment"? This project could start with the research literature or some other resource, followed by the adaptation and implementation of something for the classroom. An evaluation of the "experiment" would come with the assessment of student learning. Here, too, is a body of literature that science educators can employ. It provides information about a variety of evaluation strategies and assessment techniques to improve student learning and educate teachers about students as learners and their experiences in the classroom or laboratory.

Approaching teaching as a research activity also suggests that there are other research areas in the literature that can be utilized. Here, students become the object of one's study in much the same way that Barbara McClintock developed "a feeling for the organism" in her Nobel Prize-winning approach to the study of maize. For teachers, this means developing an understanding of how students are experiencing the classroom. This perspective would have many results. For example, one important implication for science teachers would be their need to be aware of the understanding of the scientific enterprise that students bring with them into class, since these ideas influence students' learning of science. The areas of Science and Technology Studies or Science and Technology in Society examine such questions. This literature also suggests how basic science can be given a more relevant context in the classroom with, for example, curricula centered on case studies (e.g., toxic waste in Love Canal).

The research literature can be very useful in understanding general patterns and trends in students' experiences of higher education, such as the differences that gender, race, and class make to the experiences of students in science and math courses. Here, women's studies, ethnic studies, and the research on the experience of women, girls, and minorities with and in science is important. An example of the use of this literature comes from the work of Elaine Seymour and Nancy Hewitt in their report *Talking About Leaving: Factors Contributing to High Attrition Rates Among Science, Mathematics and Engineering Undergraduate Majors*. Contrary to the common cultural explanation, the authors found no evidence that women leave science, math, and engineering majors more often than men due to a perceived less natural aptitude for these majors. Approximately the same numbers of male and female students, about 10%, leave these majors for this reason.[2] Familiarity with research such as this can offer challenges to a teacher's perception of her students and thereby suggest improvements for her practice of teaching.

The scholarship in the emerging field of Gender and Science, an area of scholarship which is at the intersection of science and technology studies and women's studies, is also useful for the teaching of science and math (see Appendix B for reading suggestions in this area). Some major areas of study in this field and their applications to teaching are: 1) the study of the history of women in scientific and technological enterprises. One important application for teaching is to illustrate that women have participated in science historically; 2) barriers to women and girls in science and ways to overcome them in the past and present. Students can be inspired from an account of the obstacles women have faced and mastered. Instructors need to be aware of the barriers that have been identified for women's and girls' participation in science; and 3) how science and technology construct gender. One example here would be the differences between the standard Biology of Man course and a Biology of Women course (such as the one pioneered by Ethel Sloane, the first Distinguished Visiting Professor in the UW Women and Science Project, who died while this book was being completed).

Finally, one narrowly reductionist approach is to understand students as brains in bodies. The areas of cognitive science and neuroscience are developing a deeper understanding of how the brain functions, and very often this can have implications for teaching and learning. As one example, Frank Betts of the Association for Supervision and Curriculum Development suggests that "our memory is very poor in rote semantic situations. It is best in contextual, even episodic-oriented situations."[3] This research indicates that students need to be introduced to concepts in a way that provides context to the concepts, rather than in a rote manner as a list of things to know. The cognitive and neurosciences can therefore provide important insights for the teaching of science, math, and engineering by providing insights into how students learn.

A teaching practice rooted in the research process and the various research literatures suggested above is a more obviously examined teaching practice than one based only on tradition. But is it sufficiently grounded? Does it provide the kind of foundation that will support innovation and adaptation to the particularities of one's local context? Does it support systemic change—a goal of the UW project? Is it a practical route to revitalizing teaching? A major problem encountered with this approach is the magnitude of the task facing a narrowly trained disciplinary specialist. It is unrealistic to expect that she or he will be able to master the literature on teaching and learning, cultural studies, and cognitive and neuroscience for a reflective teaching practice and simultaneously conduct their faculty roles of teaching, scholarly activity, committee work, advising, and so forth.

Examining what is at the core of a reflective teaching practice provides a more direct route to establishing a grounded version of it. Such a practice acknowledges that teaching itself can be a research activity and therefore potentially a collaborative activity as well. It is an ongoing and dynamic procedure whereby the research process, based in classroom experiences, is used to develop classroom activities. This is followed by implementation, observations, and evaluations of the activities, and the cycle is completed with the development of appropriate explanations for these observations which suggest how to further develop the activity, usually in consultation with others—a community of scholars. Here the teacher, like the scientist, builds explanations for the observations she is making about teaching. This kind of reflective model of teaching is really challenging how we know what we know about teaching. Challenges such as this are epistemological in nature. This focus on the epistemology of teaching leads to the understanding that how we teach—our pedagogy—is rooted in what we believe about teaching and learning and the knowledge we are teaching. Perhaps, then, pedagogy can be firmly grounded in epistemology.

This suggests that the process of revitalizing teaching by developing reflective teaching practices, of making deeply rooted changes in teaching, could then be driven by a reexamination of our understanding of our beliefs about knowledge. This in turn indicates that teachers, in addition to being disciplinary experts, also need to develop an explicit understanding of epistemology. With such an understanding, an instructor has the foundation to examine the various research literatures and other sources of ideas and make appropriate adaptations to her local contexts, to innovate their teaching, and then reflect on it. What follows is an introduction to the epistemology of science and science teaching from perspective of a scientist and a practicing science educator.

## Grounding Pedagogy in Epistemology

Beliefs about knowledge in science can be characterized by a simple model. If knowledge beliefs were placed on a continuum, one end would represent *positivism* (realism), which asserts that there is an independent body of knowledge about the natural world that exists outside of and independent of the human mind. Modern science has its roots in positivism, as Evelyn Fox Keller has described. Francis Bacon, the "father" of science, used the metaphor "lift the veil of nature and she will reveal herself to us" for the activity of the scientist: The scientist out in nature uncovers knowledge and truths about the natural world. The science student, in the lab, then verifies what scientists have discovered in nature.

On the other end of the scale I would place *constructivism* (antirealism), which has two basic versions, as defined by M. Matthews in his book *Teaching Science: The Role of History and Philosophy of Science*.[4] In the psychological version, there are two types of constructivism: in the first, based on Piaget's analysis, individuals construct knowledge from their activity in the world, in order to explain it; in the second type, from Vygotsky's work, groups or language communities construct knowledge. According to Matthews, "It ignores the individual psychological mechanisms of belief construction and focuses on the extraindividual social circumstances that determines the beliefs of individuals." Here, the scientist out in nature makes observations and proceeds to construct or invent explanations (alone, in community, or some combination of these) that work well to explain these observations and are consistent with and acceptable to the scientific community. The implication for the science student in the laboratory is that a lab exercise becomes a matter of making observations and then using the concepts from the scientific community, learning to develop explanations that work well to explain these observations.

These two extreme positions on beliefs about knowledge have an impact on how teachers teach. Current catch phrases "sage on the stage" using the "chalk and talk method" versus the "guide by the side" approach illustrate how teaching styles are related to knowledge beliefs. Many of us were trained in the "sage on the stage" mode, where the teacher delivers knowledge that scientists obtained from nature and that exists independent of humans. In contrast, the teacher-as-guide model is rooted in constructivism. The teacher guides the students' conceptual understanding in order to make sense of observations they have made. Making a switch to this constructivist position in our teaching style requires that we understand our own epistemological history.

Most of us are "recovering positivists" and this is where we run into trouble as we begin to reorient our teaching. We often attempt constructivist, guided inquiry and discovery-based hands-on activities out of our positivist beliefs about knowledge. With this inconsistency between what we were trained to believe about knowledge and the desire to have students investigate, we are essentially saying to students, "I want you to explore, but you have to find what I want you to find and I am not going to tell you what that is before you start." Then we are disappointed when students don't exhibit an effort to explore. Constructivist exploration is crafted so that most students do find what you want them to find, but it also acknowledges that students may find other interesting things, as well as accepting that the process of exploring is a key feature of the activity. Ideally, for effective teaching, we need to strive for a pedagogy that is epistemologically consistent. However, given the constraints teachers find themselves facing, perhaps the best we can hope for is that we can learn to use different teaching styles with more epistemological consistency and hence more appropriately.

There is another issue that is important to consider as we build a pedagogy consistent with our epistemology. Cathleen Loving, in her work on the preparation of science teachers, suggests that it is also important for teachers to consider beliefs about the relationship of knowers to knowledge.[5] She expresses this aspect of epistemology as a continuum from a rational position to a natural position. With the extreme rational position goes the belief that humans have the capacity to be purely objective, with the appropriate methodology, to "step outside themselves," to go out into nature, make objective observations, and then discover objective explanations. The other extreme position, the "natural" position, says that humans are products of their culture and their gender, race, and class socialization, and are unable to get outside of their humanness. Observations are culturally laden and biologically limited—we see what we are positioned to see—and our explanations are laden as well. I overemphasize the extremes to make the point that in science, the actual practices probably lie somewhere between the two ends of this continuum.

These beliefs about the relationship between knowers and knowledge have important implications for learners, as they address the question of who can know. Since many students have taken on the belief that scientists are objective in the purest sense, they forget that what they are learning was developed by humans who are scientists. Textbooks and the knowledge therein seem to them to be devoid of any human connection. For many students, understanding that what they are attempting to learn comes to them through the work of others—scientists—makes it seem more possible that they, too, can acquire this knowledge. At the other extreme, there are some students who have adopted a cultural belief that all statistical data (and by implication, scientific knowledge) is hopelessly biased. Learning science then becomes oppressive. These students may need to come to the understanding that science, while practiced by humans, and despite its limitations, does offer a reliable method for developing explanations for our observations of the natural world. Explicit discussions of this aspect of epistemology are critical for students as they examine their motivation for learning science.

This short description of the epistemology of science, with a few examples suggesting the implications for the teaching and learning of science, illustrates the importance for teachers and students of developing an understanding of epistemology and its relationship to pedagogy. Cathleen Loving offers a valuable perspective on this process of exploring these issues. She has charted on an x,y graph the two epistemological beliefs discussed above—the nature of scientific knowledge and the relationship between that knowledge and knowers—of prominent philosophers of science to show us that there is no agreement among them about the epistemological foundations of science. Understanding that there is a lack of agreement allows us some flexibility as we explore these issues for ourselves as scientists and science educators. Furthermore, for science educators concerned about equity issues in science and science education, the fact that many of the feminist critiques of science are fundamentally concerned with epistemological issues is important. Epistemology is intimately linked with teaching and learning—and therefore feminist critiques and reforms in the sciences are also linked to teaching and learning.

As we work toward crafting a reflective teaching practice that is inclusive and grounded in epistemology, it is important to consider the scope of the epistemological beliefs to be examined. The work of Nona Lyons suggests that for teachers, a delineation of three types of epistemological beliefs ought to be considered. These are the teacher's beliefs about knowledge, the teacher's understanding of her students' epistemological beliefs, and the teacher's understanding of the discipline's epistemological foundation.[6] As I have already mentioned, one step toward understanding our own beliefs about knowledge and the foundation of our disciplines is to explore the beliefs that students hold. The work of Aikenhead and Ryan provides a model for an instrument that can be used in evaluating students' beliefs about knowledge.[7]

Many introductory courses and texts begin with a very traditional, often unexamined presentation of "The Scientific Method," which is intended as the students' introduction to the epistemology of science. Here is a good place to take the time to evaluate students' understanding of epistemology and to explore the topic with them in more depth. Specific discussions of epistemology can have the effect of increasing students' interest in science. One reason this is so is that it provides an opportunity to explore with them how issues of gender, race, and class impact knowledge and knowers—to explore the human face of science. Consider the impact of gender on the two epistemological issues presented above: the nature of knowledge and the relationship of knowers to that knowledge. Historically, positivist science knowledge has been gendered; traditionally, only men, the rational beings (or so the biologists of the past have asserted), have had the ability to "lift the veil of nature." By contrast, accepting scientific knowledge as constructivist suggests that both men and women can construct it. Furthermore, the differences among perspectives (between men, between women, and between genders) is welcomed. A constructivist science is an enriched science, since perspective makes a difference in what scientists choose to study, how they study, and how they explain their observations. Many female students have commented to me on the importance to them of understanding these aspects of epistemology. Some report that, for them, a science rooted in constructivism gives them permission to participate fully and succeed in science.

This brief discussion of epistemology has raised a key issue: beliefs about knowledge are central to teaching and learning. In order for teaching to be revitalized so that it is more inclusive as well as more effective, these beliefs must be explored and examined and their relationship to teaching and learning made explicit.

## Constructivism as a Foundation for Pedagogy

Remember the conundrum that initiated my investigation into the foundations of teaching? Students in the early courses I taught seemed to be understanding concepts better, but they were not learning how to think critically about these concepts. They did not comprehend that the discipline of chemistry (or any other one) provides a distinct perspective, a way of thinking about and exploring the material world. Furthermore, there were other things that I hoped my students would do but never seemed to. They did not appear to be learning how to be investigative, and no spirit of tinkering or urge to inquire seemed to be developing. They also appeared hopelessly confused about the relationship between observation and theory. The kind of teaching I had been pursuing, based in an epistemological tradition of realism, seems to have as its main goal the development of a student's knowledge base within a discipline. While this is an important activity, it does not accomplish all that we hope an education in science might provide. Teaching based in constructivism offers a wider array of outcomes for learners, outcomes that include the development of a discipline-specific knowledge base.

A comparison of different teaching practices in laboratory courses illustrates these different epistemological positions at work in teaching. From the positivist tradition, in many introductory science labs, the primary learning activity is verifying scientific concepts. The lab begins with a prelab lecture where the student is told what observations to make, how to make them safely, and what the explanation is for their observations (which they have not yet made). Students are often led through the lab with a fill-in-the-blank worksheet, where they record observations that are often graded for correctness. At the end of the lab, on their own time, the student is often required to carry out the specified calculations and produce a professional x,y graph and interpret it to match the explanation they were given in their prelab lecture. It is assumed that through this exercise, students will not only learn technique and methodology, but they will also develop a conceptual understanding of the material.

An introductory science lab cast in the constructivist tradition is decidedly different. The primary learning activity is one of inquiry or discovery, either guided by the instructor or an open one, but in both cases with the exploration moving from observations to explanations. A prelab lecture consists of an introduction to the system that is the focus of the lab, including the presentation of safety considerations. Here students' prior experiences with the system and their understanding of it are elicited. Then students, often in teams, are given a question for investigation with some guidance on how to proceed. After teams have made observations, they are brought back together to share data and discuss what needs to be done in order to develop an explanation (via calculations and graphs). This kind of lab activity makes the process of investigation central, illustrates the relationship between observation and explanation, and thereby enhances students' conceptual understanding and critical thinking.

I have only presented one example from a laboratory course to illustrate how teaching and learning rooted in constructivism differ from more traditional methods. These differences can be seen in classroom practices as well. The hallmark of these constructivist practices is the recognition that students, as knowers, construct their understanding based on prior conceptions and experiences, and on the experiences presented to them in the classroom setting. These experiences, crafted by the teacher, guide students in developing understandings that are in accordance with those of the discipline. Then, as an evaluation step, students are asked to apply their new or modified understanding to a different problem or situation.[8]

An important point needs to be emphasized here about constructivist classroom practice, particularly as it relates to teaching and learning in the sciences: Constructivism asserts that all students—male and female, majority and minority—can know, and that their knowing or coming to know is influenced by their classroom experiences and their prior experiences, which are in turn shaped by the specifics of their lives, but also by the cultures in which they live. It follows, then, that to be effective teachers, constructivist teachers must be informed about how issues of race, class, and gender—their own and their students'—influence teaching and learning in the constructivist mode.

How do teachers who have been teaching as they were taught begin teaching in this constructivist mode? In the following chapter, examples of project participants' attempts to change their pedagogy are presented. They demonstrate how faculty members learn to alter their teaching styles and bring to the classroom pedagogical practices grounded in constructivism. At the core of these innovations are changes in fundamental beliefs about knowledge, learning, and teaching. In addition to changes in pedagogy, raising epistemological questions also raises questions about two other important aspects of teaching: 1) the content of courses and the design of curriculum, and 2) the atmosphere or climate of the class for teachers and students. These three elements—pedagogy, content, and climate—cannot readily be isolated from one another in the work of revitalizing teaching, as they all have roots in foundational beliefs about knowledge, teaching and learning.

## Constructivism, Course Content and Classroom Climate

In most departments, the curriculum for science and math majors as well as the content of the courses in that curriculum is hotly contested territory. Many science, math, and engineering courses are departmentally owned and operated. That is, in multisection courses there are often common learning outcomes and even common syllabi and finals. In addition to certain courses, the curriculum for a major as a whole is departmentally dictated, based on departmental understandings of the discipline. Graduation requirements must also be considered, and many of these courses and curricula come under the scrutiny of accrediting agencies and professional societies. These limitations present interesting problems for those actively revitalizing their teaching and addressing issues of gender in science. They point to the far-reaching implications of epistemology on the practice of teaching and learning.

What is meant by "content" changes with constructivist approaches to teaching. Content includes the knowledge base and the process skills of the discipline, as well as learning about learning and other epistemological issues. The limitations placed on curriculum and course material present a significant barrier to the implementation of different notions of content, but there are some strategies that can be used to overcome these barriers and thereby create course syllabi more consistent with the goals of student-centered learning. For instance, it may be possible in some circumstances to reorganize the order in which course content is introduced in order to enhance students' understanding. Bringing real-world applications and case studies into a course is another change in content that can be readily made and which increases the relevance of a course for a student, especially those from applied majors. Beyond these relatively simple changes, faculty must engage in the long-term political process of developing new courses and transforming course outlines and entire curricula. (Chapter 4 provides additional examples of changing course content within the constraints that bind curriculum.)

Another important area of focus in the constructivist classroom is classroom atmosphere, or climate. This refers to how students relate to each other and to the instructor in terms of teaching and learning. A shift to constructivism—which acknowledges that individuals, with their particular learning styles, in groups or alone, construct knowledge—requires that attention be paid to the classroom climate. A constructivist learning environment is one where students are working together with the instructor to learn the course material and the instructor is committed to a reflective teaching process. Designing course structure and pedagogy that supports the building of these kinds of learning communities is a significant task. Attention must be paid to individuals as learners as well as to collaborative and cooperative activities for learning. Journal writing and other assessment activities can aid in the monitoring and development of a classroom atmosphere that invites collaboration. One critical aspect in the development of these collaborative communities is that the individuals within groups must understand their interdependence. Carefully crafted assignments need to be designed with this goal in mind. The considerable research on collaborative and cooperative learning, learning styles, and the impact of gender, race, and class on education can be utilized in the development of classroom climate appropriate for learning. (Chapter 4 offers further examples of the activities of faculty addressing climate issues.)

This has been a very general description of how epistemology impacts course content and classroom climate, and particularly how a constructivist approach to pedagogy calls these aspects of teaching into question. Chapters 3 and 4 will illustrate specific examples of how faculty participants in the University of Wisconsin Women and Science Project grappled with changes in pedagogy, content, and climate as they explored epistemological issues in science and math, as well as issues of gender. These faculty members were challenged to make changes in their teaching, a challenge that also placed them in the situation of being learners—learners about teaching. Can we apply what we have learned in the preceding sections about students as learners to faculty as learners as they participate in faculty development activities?

## Constructivist Faculty Development

Since teachers generally "teach as they are taught," and since faculty are being asked to teach in different ways to distinct student populations, a constructivist approach also requires faculty development activities that are designed specifically to provide them with opportunities to examine the epistemological foundations of teaching. This focus on epistemology—as we have seen in the previous section, where it was applied to students—leads into an inquiry of how gender, race, and class impact knowledge and knowledge production. For many faculty members, this epistemological connection between gender and knowledge stimulates them to reevaluate their own pedagogy. To be effective, this kind of foundational examination not only requires that faculty think deeply about teaching, but in addition, they must experience the type of instruction that they are being asked to develop. In other words, for faculty to develop pedagogy that is constructivist, constructivist faculty development activities need to be made available to them. Importantly, these activities must be planned to give attention to the process of change over time.

Explicitly constructivist faculty development is in contrast to more traditional faculty development activities. These conventional kinds of programs seem to have as their primary goal the delivery of information, in a timely manner, about topics such as teaching methods, learning styles, course activities, and educational technology. Activities such as these have as their foundation a particular epistemological position regarding teaching: that knowledge about teaching is received and not constructed. While this kind of activity does make a contribution to the development of teaching practices, it is not effective for the transformation from a traditional pedagogy to a constructivist one. Faculty development, like teaching, must be epistemologically consistent.

## Faculty Development in the Women and Science Program

The design of the original University of Wisconsin System Women and Science project exemplifies a constructivist faculty development process, as shown by several key features. It was a long-term project that supported the goal of fostering systemic change in order to develop more effective and more inclusive teaching in science and math. The intentional creation of learning communities of faculty focused on the pedagogical, climate, and content issues related to teaching and learning is central to sustaining the faculty development. In this particular project, these communities were supported by local on-campus activities, workshops with nearby campuses, and an annual system-wide Women and Science retreat. The resources for learning in these communities of faculty were initially provided through the activities led by Distinguished Visiting Professors (DVPs). These leaders, located on campuses for at least one semester, modeled constructivist strategies cast in a feminist perspective as "guides by the side" rather than "sages on the stage." Many of the activities they conducted were created to provide faculty participants with the opportunity to experience constructivist teaching and learning firsthand.

Not surprisingly, faculty participants were very sophisticated learners, and the value of development activities of this kind often became apparent only after reflection on their participation in them. Many reported that they eventually realized that what was critical to them about their participation in the project was that they were learning about teaching, often in small groups, in much the same way they were beginning to ask their students to learn. They repeatedly suggested that this faculty development experience, with its emphasis on learning communities, was reminiscent of their graduate student days, and it can in fact be argued that graduate education as it is commonly practiced in science and mathematics is epistemologically grounded in constructivism. It is this experience of learning at the graduate level that many faculty members would like to see incorporated, even at an elementary level, in undergraduate education.

## Distinguished Visiting Professor Project Activities

One of the strengths of the Distinguished Visiting Professor model of faculty development characteristic of the Women and Science Project was the presence of a resident guide for faculty development. While in residency in a science or math department, the DVP had several responsibilities, including: 1) modeling teaching; 2) serving as a resource on issues of pedagogy, content, and climate generally and more specifically as related to gender issues; 3) leading discussions and stimulating conversation and reflection about these issues; 4) contributing discipline-specific expertise; and 5) making presentations as a guest speaker to seminars, courses, student groups, and at other campus events. Significantly, one of the underlying goals of all these activities was to foster a community of faculty that would continue the project long after the end of the DVP's residence.

As a faculty development leader, it was crucial that the DVP intentionally, consistently, and continually model constructivist strategies. This is the hallmark of a constructivist classroom as well, in which the teacher models the process of constructing knowledge. DVPs achieved this by teaching one-semester courses that were open to the public; that is, faculty formally participating in the Women and Science project (Faculty Fellows), as well as other faculty from the campus, were invited to drop in and observe. Since little departmental discussion usually focuses on teaching issues, these visits provided faculty with a much-needed opportunity to discuss issues centered around teaching and learning with each other and with the DVP. Tandem teaching was another very effective vehicle for promoting constructivist learning about pedagogy, content, and climate issues. Here, the DVP and a Faculty Fellow taught different sections of the same course, an activity which often stimulated plenty of conversation about course issues, and hence learning about teaching, on the part of the Faculty Fellow and the DVP.

Conducting workshops was another important aspect of Distinguished Visiting Professor leadership in the Women and Science project. DVPs designed workshops with a constructive foundation that they then implemented with Faculty Fellows. These workshops often began with a plenary presentation by the DVP, followed by sessions presented by the Faculty Fellows. These sessions gave Faculty Fellows the opportunity to construct new knowledge on teaching and learning techniques by requiring them to develop presentations collaboratively for faculty colleagues attending the workshop. (This is a clear example of a constructivist strategy for faculty development, devised by one of the first DVPs in the project, Sherrie Nicol.) The process for developing these sessions began with the DVP presenting Faculty Fellows a list of potential topics on teaching and learning for them to explore. Fellows selected topics they were interested in learning more about, an important feature of any constructivist lesson. The DVP guided the faculty in the preparation of their sessions, providing resources, along with examples of how to use those resources. A key feature of this process was the multiple discussion sessions with faculty (sometimes by email), led by the DVP, who directed them in the development of their topics. (Session topics included classroom assessment techniques, mentoring female students, learning cycle activities, collaborative journalling, and critical thinking). The process by which Faculty Fellows prepared a presentation was very much like some of the small-group activities that have been developed for students—an observation that Faculty Fellows usually made sometime after their first workshop presentation.

Plenary sessions given by the Distinguished Visiting Professor were another important feature of the faculty development workshops. In addition to challenging fundamental epistemological notions about teaching and learning, these sessions also provided faculty participants with a framework for thinking about their teaching as a reflective practice, once again with an emphasis on pedagogy, content, and climate aspects of teaching. In order to develop such a practice, it is necessary for many faculty members, trained as disciplinary experts, to be introduced to research-based resources on these aspects of teaching in science and math education. In the UW project, the DVP cast this introduction to resources in a constructivist mode—modeling their use and providing experiences for faculty to develop or apply these resources to their own teaching situations, often in different disciplines of science or math from the DVPs.

This cross-disciplinary conversation was a significant element in the establishment of learning communities of faculty focused on developing reflective teaching practices. (Many other activities that took place in the original Women and Science Project were integral to the formation of learning communities of faculty, as well as to their ongoing development. These will be described in later chapters, where much more detail will be provided from the multiple perspectives of Faculty Fellows, DVPs, campus administrators, and the Director of the now-permanent UW System Women and Science Program.)

This brief discussion of faculty development activities, along with an introduction to the Distinguished Visiting Professor model of faculty development central to the Women and Science Project, underscores two major points. First, there is an underlying, often unexamined and unstated, epistemological foundation to faculty development activities, just as there is to teaching. Second, there was a particular epistemological foundation in the UW System Women and Science project, a constructivist one that cast faculty in the role of learners experiencing the kind of learning activities they were being asked to develop for their students. The project's unique focus on gender issues in pedagogy, content, and climate was intimately related to the emphasis on these foundational issues.

## Conclusion

A systemic revitalization of teaching requires a reexamination of core beliefs about teaching, learning, and faculty development. For teaching to be effective for today's learners, instructors must have a strong understanding of their own disciplines as well as an explicit understanding of epistemology, since both are critical basic elements of a reflective teaching practice. In such a practice, courses must be designed with attention to building pedagogy, classroom climate, and content that are epistemologically consistent. Faculty must receive support in the development of such courses through intentional faculty development projects that are also epistemologically consistent. Institutions must be prepared for the formidable task of meeting the challenges that these different approaches to teaching as well as to faculty development create.

The next chapter explores some of the pedagogical changes that faculty undertook as a result of their participation in the Women and Science Project. On one level it can be read for examples of innovations in teaching and their roots in epistemology. On another level, though, it is the story of faculty undergoing a process of change—challenging deeply held beliefs about teaching and learning and about how they develop as teachers.

## NOTES

1. S. Leiwan, "Four Teacher-Friendly Postulates for Surviving in a Sea of Change," *The Mathematics Teacher* 87 #8 (1994): 392-393.

2. E. Seymour and N. M. Hewitt, *Talking about Leaving: Factors Contributing to High Attrition Rates among Science, Mathematics, and Engineering Undergraduate Majors*, Final Report to the Alfred P. Sloan Foundation on Ethnographic Inquiry at Seven Institutions, Bureau of Sociological Research, University of Colorado, Boulder, 1994.

3. F. Betts, "Cognitive Research and Its Implications for the Classroom," Plenary Session, La Crosse Public Schools, April 1995.

4. M. Matthews, *Teaching Science: The Role of History and Philosophy of Science* (New York: Routledge, 1994).

5. C. C. Loving, "The Scientific Theory Profile: A Philosophy of Science Model for Science Teachers," *Journal of Research in Science Teaching* 28 #9 (1991): 823-838.

6. N. Lyons, "Dimensions of Knowing: Ethical and Epistemological Dimensions of Teachers' Work and Development," in L. Stone, ed., *The Education Feminism Reader* (New York: Routledge, 1994).

7. G. S. Aikenhead and A. G. Ryan, "The Development of a New Instrument: Views on Science-Technology-Science," *Science Education* 76 #5 (1992): 477-491.

8. Kenneth Tobin, ed., *The Practice of Constructivism in Science Education* (Hillsdale, NJ: Lawrence Erlbaum, 1995).

*Chapter 3*

# Transforming Pedagogy

**Laura Stempel with Cheryl Ney and Jacqueline Ross**

> We must help students see themselves as part of the community of science. . . . Women and minorities are the miner's canary signaling deeper problems in our programs. We must recognize that as the demographics change we are playing to a "tougher house." If we do nothing to rethink our programs and depend on old strategies of weeding we face a troubled future. For not only are we not drawing proportionately from the disenfranchised majority, we are losing the interest of traditional participants as well.
>
> —Shirley Malcolm, Director, Education and Human Resources Programs, American Association for the Advancement of Science

In an influential 1992 essay entitled "Science in a Postmodern World" that addressed the recent decline in the number of college students majoring in the sciences and going on to pursue scientific careers, Kenneth Bruffee described the gap between what scientists actually do and what they teach:

> The heart of the problem is the tension between the way scientists do science and the way they tend to teach science. Scientists as scientists [follow] a tradition based on the interpretive ability, in collaboration with other scientists, to construct, manipulate, and calibrate models and symbol systems. As teachers, however, scientists present themselves . . . as . . . something a little like museum curators, as if a scientist's main job were to accumulate, maintain, and display curious and useful facts about the natural world.[1]

The University of Wisconsin Women's Studies Consortium's Women and Science Project was part of the attempt to bridge this gap between what scientists actually do and what they teach students that science is. One goal of the project and the permanent program that guild on it has been to make science, math, computer science, engineering, and related fields more accessible to students who might otherwise be intimidated or simply unmoved by these disciplines—including women and members of other underrepresented groups. But even more important has been the hope of improving students' conceptual understandings and thinking skills, and thereby enhancing their interest in technical and scientific careers. Sheila Tobias, whose germinal work in this area helped to shape the project's original parameters, puts it this way: "To deal with the projected shortfall [of students trained in the sciences], we are obliged to think and think hard not just about who does science and why, but who doesn't do science and why."[2]

This chapter examines the changes in classroom practices that have been at the center of many project participants' efforts. While faculty development activities such as workshops focused on building collaborative communities among interested teachers throughout the University of Wisconsin System, the ultimate purpose of these communities was directed at the students themselves. The aim, after all, was to alter and enhance the participants' pedagogical methods in order to retain promising students who would otherwise be likely to abandon the study of science as soon as their general education requirements were satisfied. (We return to this topic in Chapter 4.) The project thus promoted pedagogies that place students' interests and needs at the center of the curriculum and that encourage them to take more active responsibility for their own learning. This chapter proceeds from relatively simple alterations in course structure and assignments to more fundamental and complex ones, in much the way a newcomer to the project of change might proceed. Because the practices described here help to increase students' actual understandings of fundamental scientific concepts, students who are not already interested in the subjects themselves are likely to become so, and thus perhaps to go on to major in the sciences and even to enter a scientific or technical career.

Faculty Fellows, Distinguished Visiting Professors (DVPs), and faculty workshop attendees experimented with a variety of pedagogical practices both during and after their formal tenure in the project. Some techniques were relatively simple and straightforward—for example, using short papers to explore specific scientific concepts—while others, such as shifting to a "discovery" approach (in which students work out a concept from evidence, rather than having the concept laid out for them) involved fundamental redefinitions of how science is learned and taught. Although this chapter describes the experiences of individual project participants using a variety of specific pedagogical practices, it is not meant as a straightforward guidebook of advice. Instead, the idea is to suggest a range of possibilities for changing the college science classroom in order to challenge students to become more active learners. What the various approaches below share is the attempt to move from the dominance of teacher-centered approaches, such as lecturing, to pedagogies that focus on the students themselves. It's worth noting that, while many of the methods used in the Women and Science project were specifically devised in response to studies of how women learn, the major pedagogical techniques—such as teaching students to consider the relation of individual pieces of information to a larger context or even to their own professional development—have proven to be equally effective for men.

Most of these classroom practices have been in use for many years, particularly in interdisciplinary fields like Women's Studies, and some science instructors have long been in the habit of designing their courses around student-oriented activities. What has been innovative about the Women and Science project is its holistic approach and a constructivist base informed by feminism—both of which emphasize a fundamental rethinking of science pedagogy—and the fact that the project has involved a collaborative effort among many faculty members, rather than isolated teachers experimenting in their own individual classrooms. Significantly, that collaboration among faculty echoes the collaborations students are encouraged to develop as well. As project participants' reports on their teaching experiences demonstrate, individual faculty members had varying success with the techniques they used, and this apparently depended in part on their own prior pedagogical assumptions and experience. For some, the approaches they tried were completely new and forced them to question their own long-standing premises about how science, math, engineering and computing should and can be taught. Others had employed some of these methods into earlier courses, and saw this project as an opportunity to demonstrate their effectiveness to faculty colleagues and to reform the broader science curriculum so as to incorporate innovations they already believed in. In almost every case, faculty members found that change was extremely time-consuming, requiring extra preparation, more review of student work, and greater attention to the classroom atmosphere. Regardless of the specific subjects they were teaching, the approaches they chose, and their own prior experience, all of the participants had to contend with a range of student attitudes and responses to changes in classroom practice—an important reminder that students as well as teachers come into the classroom with a pre-existing set of ideas about how learning should proceed. Yet most project participants concluded that both they and their students benefited from the changes they instituted.

## A Self-Monitoring Exercise

One of the most basic problems faculty members in all disciplines face, regardless of their attitude toward the existing curriculum, is the difficulty of assessing both their own performance and the depth of students' understanding of the course material. Before content or pedagogy can be adapted to students' needs, teachers must be able to evaluate the effectiveness—or ineffectiveness—of established and familiar practices, and such assessment must obviously continue as change is being implemented. "Too often," writes UW-Stout Faculty Fellow Janice Gehrke,

> we assume that no one could possible fail to follow our brilliantly devised, computer generated, color-coded diagrams, and that students will be eager to ask questions about anything that they don't understand. [But] we are regularly faced with disappointing evidence when we look at papers, quizzes, and tests. . . . There are gaps, yea, even light years, between what we thought we were teaching and what they thought they were learning. Furthermore, by the time we notice these gaps in knowledge or understanding or both, it is frequently too late to remedy the problems.

For Gehrke, who teaches biology, student self-monitoring is a key to bridging those gaps. This is especially important, she believes, when it comes to sorting significant from insignificant information, a necessary first step in answering questions she sees as basic to learning scientific concepts, such as how a particular piece of information relates to the entire subject of the course. "One helpful sorting technique," writes Gehrke, "is the minute paper, in which students anonymously write down everything they can think of concerning a particular subject, organism, or process in one minute. Papers are then exchanged and several are read aloud by other students [who] jot down a list of those items mentioned most often." Students then discuss the individual items' significance, with the instructor suggesting possible answers if discussion lags (rather than simply giving them the explanation, as might happen in a more conventional classroom).

Gehrke notes that students also "need help in exercising their imagination. Often they are well trained in the art of memorization, [but] have had precious little opportunity to sketch ideas, make up dramas, or fashion the rules of the game." She uses sketches and diagrams to help them think about what she calls the "shape of a particular piece of information," and like many other participants in the project, she also emphasizes the notion of *relatedness*—in this case, among particular pieces of information.

Gehrke has discovered that students are generally positive in their response to these new approaches, and that they find the use of worksheets, for example, helpful in understanding the connections among ideas and pieces of information that might otherwise seem unrelated. Because developing this sort of teaching plan takes a good deal of time, she suggests that faculty members begin with a course in which they have a lot of experience and expertise. And like many project Fellows, she cautions that teachers must reevaluate some of their most basic classroom goals: "A lot of people have asked, 'Can you cover as much material?' The answer is 'No.' Do we need to cover as much material? Probably not. We need to focus on *how* our students are learning rather than on how much material we are covering."

## Diversified Instruction

For Andrew Balas, who teaches mathematics at the UW-Eau Claire campus, an experiment in diversified instruction suggests that changes in classroom practice can have important benefits for both students and faculty members. His experience is worth examining in some detail, because it illustrates both the advantages of changing the pedagogical methods used in traditional courses and the difficulties faculty members face when they experiment in this way.

By taking two different approaches to sections of an algebra course, Balas was able to compare his own and his students' reactions to the use of practices to which he was introduced in the Women and Science project. Although he did not conduct a strictly controlled experiment, Balas studied the differences between two sections, one of which he designated as the "control." In that section, he used the approaches he had employed in previous years: answering student questions, working homework problems on the blackboard, explaining new material, and giving conventional quizzes and tests throughout the semester. In the so-called "experimental" section, however, he drew on a variety of methods that represented his attempts "to induce the students to use themselves and each other as authorities, not me."

For instance, rather than reviewing the homework questions on the blackboard and providing the solutions himself, Balas divided the experimental class into small groups that went over homework assignments together. "If there was a problem that no one could do," he reports, "they would inform me and I would find a student from another group to put the problem on the board." Students would also work on new material in their groups, using worksheets based on the discovery method, and even taking some of the semester's quizzes together. Although several of the course exams were the same as those given in the "control" section, some took a different format that allowed students to choose which segments to answer. For one test, Balas even "sold hints" to difficult questions, deducting points from students' scores depending on how much of a problem he worked for them. Those in the experimental section also had the option of writing a journal for extra credit and of replacing half their final exam grade with a paper or other project.

Balas' semester was a stressful one, partly because of his own lack of experience with these new teaching methods. Some of the small groups were unsuccessful, often because they lacked students with a high level of competence and therefore were unable to complete homework assignments or perform well on tests and quizzes. At least one student complained publicly that Balas was neglecting his duty by not lecturing, and daily classroom interaction reinforced his sense that morale was low. As the semester went on, "the atmosphere in class deteriorated. I came to dread meetings with the experimental class. . . . I felt inadequate to the task." Knowing that "resistance is a self-fulfilling prophecy [and] students will have difficulty learning if they do not feel comfortable with the teaching methods," Balas decided to poll the class at midterm. He thus discovered that only a small percentage of students had fundamental problems with his methods, and that most were actually enthusiastic about the diversified approach. (Interestingly, he noticed that "the strongest reactions to these teaching methods—favorable as well as unfavorable—came from women.") Still, he did begin to lecture more often, instituted an attendance policy, and offered new grading options. The atmosphere changed for the better and "once again it became a pleasure to teach the class."

The composition of these two sections differed: members of the "control" section had higher pre-test scores and many more planned to take additional math courses. But Balas' analysis of end-of-semester evaluation forms he distributed suggests that the diversified approach had a clear impact on students' performances, and even on their responses to his pedagogical experiment. Some of those in the experimental section found the discovery and small-group methods an inefficient use of classroom time, but many more said they were highly challenged. Balas was, however, reluctant to jump to conclusions, wondering whether these differences occurred "because [the students] entered with math skills at a lower level, or because the varied instructional techniques made them think about the concepts."

Many more students in the experimental than the "control" section approved of his grading policy, "evidently [sharing] my view that the alternative assessments work in the students' favor." Their conviction that he treated them with respect once again prompted Balas to consider competing explanations: "Was this because the student-based learning elevates the students and builds their self-esteem, or was this simply a reaction to my polling of the class and making adjustments based on their recommendations?" He found, too, that students used their small groups outside of class, meeting to study and keeping each other up on missed classroom lessons. "In short, members were taking responsibility for the entire group's learning, one of the desired outcomes of small group learning." By the end of the semester, Balas also felt that he had come to know his students better through their journal entries and small-group work: "I felt a part of their lives and math careers. When it came time to assign final grades, I could picture each student as I worked out their grade."

## JOURNALLING

As this example suggests, the use of journals was an important activity for many project members. This practice has become common in the humanities and even in some social science fields, but science, math, engineering, and computer science instructors rarely employ it. Those project members who incorporated journalling into their syllabus invariably found that it helped both students and teacher, especially when it came to assessing students' experience of the courses. Rather than simply trying to read students' facial expressions during class sessions, they could read what students actually thought about how the course was going. Early in the term, for instance, both Sandra Madison (UW-Stevens Point) and Sherrie Nicol (UW-Platteville) asked students to write brief autobiographies of their computing and math experiences, respectively. This assignment allowed each instructor to get to know their students quickly and to be aware of their fears about the coursework to come. Madison also used ungraded journal entries as a way of identifying misunderstandings about specific concepts. "I invited the authors whose answers seemed bizarre to talk with me one-on-one, hoping that we could clear up the faulty conceptions before the students struggled with traded programming assignments or examinations."

Sherrie Nicol designed a wide variety of journal projects for her math students, asking them to react to the course structure, evaluate the strengths and weaknesses of their small groups, and in one case, "detail every thought they had in attempting to solve a difficult problem." Like many of the pedagogical practices instituted by project Fellows and DVPs, journal-writing requires extra work on the instructor's part. Yet Nicol is convinced of the advantages these activities offer to nearly all students, whether because they receive "an easy 'A' for a small percentage of their grade" or because the opportunity for self-reflection helps them to understand their own learning processes. Equally important, the activity also gives Nicol "a constant source of self-evaluation," allowing her to tailor classroom lessons to students' needs:

> The first week I have two required writings. The first is their personal automathography, a history of their mathematical experiences, including their successes and failures, likes and dislikes, and self-perception of their progress. The second is their reaction to the course structure. I typically have students work in groups, do projects, take group quizzes, keep a journal and use student-exploration rather than lecture. Since much of this is alien to mathematics courses, I ask the students how they feel about the way the course will run. One of their final journal entries involves looking back at this entry and reflecting on what they initially were apprehensive about.

Nicol believes that the self-reflection involved in the journal-writing process is beneficial. Still, she notes that the commitment to the journals can vary widely among the students, as does the quality of their entries: Although other project participants found that their male students wrote extremely thoughtful journal entries, Nicol notes that "my female students are far more reflective and honest in their journals, and . . . those students with good writing skills also have more meaningful journals."

The observation that students use their journals differently is also of interest to Distinguished Visiting Professor Cheryl Ney:

> The more I use journalling and writing assignments to ask students to reflect on their learning in [a] collaborative way with me, as I write back to them, the more I am noticing that in general, more women students in a course respond favorably to this writing. But it is also the case that some women students don't, can't or won't use writing to reflect on their learning and that many male students will, can and do write and reflect. One thing that seems to be common among students who are reflective is the degree to which they are active learners, engaged in the process of learning the course material and motivated to trust that the assignment is an important one. Students who are in the course simply to fill a requirement they feel has been imposed on them don't respond well to the assignment. Sometimes, over the course of the semester, it's the collaboration with me that wins over even some of these students.

Faculty Fellow Barbara Scheetz Nielsen also used journal assignments in a

UW-River Falls analytical chemistry course, to help students understand their own learning practices as well as to give them an opportunity to express their reactions to the coursework itself. Like many other project faculty, Nielsen provided explicit guidelines about the length, frequency, format, and grading of the entries, and formally assigned questions to be answered each week. These ranged from general issues—such as "Why am I enrolled in analytical chemistry?"—to topics more narrowly focused on the week's classroom work. Many of these questions required students to take an active role in learning by rewriting sample problems or class notes, or by a creative activity like producing an "autobiography" of a weak electrolyte to further their conceptual understanding. Broader assignments allowed students to draw connections "between what they learned in the course and what they were studying in other courses or something from their daily life."

Among Nielsen's most innovative assignments was one called "Roadmaps and Roadblocks":

> This [assignment] was designed to help the students identify how they learn and where their difficulties [in] problem solving lay. Most textbooks provide sample problems at the end of each chapter. . . . several of these problems were assigned, and the students were asked to work out roadmaps for each problem, . . . an outline of how the problem could be solved. . . . After completing the roadmap, the students wrote short statements about any difficulties they encountered in solving the problem. In this way, [they] began to pinpoint where difficulties arose and the instructor could add comments on how to proceed. In many cases the student's difficulties were recurring, and once identified, the student could proceed with greater understanding.

Nielsen notes that students did have some initial resistance to the idea of keeping a journal—which they associated with the humanities—in a science course. But their evaluations reveal that many recognized that the activity "forced them to organize their work and time and study the material regularly." On Nielsen's side, the benefits included establishing "an unspoken dialogue . . . with each student from the very first entry. This underlying communication allowed for a personal, friendly, and more open classroom atmosphere." She also found an improvement in the level of questions asked during office hours, since her early intervention in the learning process meant that students came to her office with many of their basic misconceptions already identified.

Distinguished Visiting Professor Vera Kolb used journals in an unusual way by having her graduate teaching assistants, rather than the undergraduate students themselves, record their experiences in the large General Chemistry class she taught at UW-Madison. Because the TAs provided her only link to the 345 students in the course, Kolb felt that she was teaching "vicariously" through them. "If I wanted to implement any women-friendly teaching goals in the labs and discussion sections, I had to do it via the TAs." She therefore asked them to use their journals like a research notebook, describing all of the teaching events they believed to be important and reading the entries aloud during their weekly meetings with her. By this method, Kolb discovered that her TAs had a wide range of teaching aims, not all of them compatible with her own. For instance, one TA took what she described as a "sink or swim" attitude toward his students, apparently believing "that good students really do not need teachers, and that the bad students cannot be helped in any way. He did not seem to believe that it is easier to swim if one is coached." Kolb is convinced "that the TAs do need to be trained in the teaching objectives, how to help all students to achieve the best of their potential, and how to present science to students in a friendly way. The journalling was a way for me to train them."

UW-Eau Claire mathematician Marc Goulet, an unofficial Faculty Fellow, writes about the impact of his experience in the Women and Science project:

> A primary shift in the classroom as a direct result of my involvement . . . means placing the student's previous experiences as a focal point for each lesson. . . . Previously, I had been using journals in my introductory mathematics classes for students not planning to go on in math and science. It was through these journals that I first realized the primary importance of student experiences in coming to know a subject.

Yet, like many other project participants, Goulet found the time required to design appropriate journal assignments and then to read students' entries excessive. He therefore replaced them with two or three essays a term, 300- to 500-word assignments that give students a chance to express the same kinds of reactions they wrote about in their journals. Writing of a student whose essay on frustration revealed that she'd been "taught to be afraid of mistakes," Goulet notes that she "is certainly a student whom we have lost in the math and science pipeline. Her essay did enable me to become aware of her past history and offer some encouraging words. Perhaps her daughter will feel differently towards the spirit of mathematical and scientific exploration." Another student named various mathematical functions after favorite TV characters and thus found that "the whole purpose of the function has suddenly become quite clear to me"—a revelation that Goulet believes indicates that, "if we enable students to come at a subject from their own point of view, we can witness some surprising connections and pleasant consequences."

## THE LEARNING CYCLE

Several project participants experimented with teaching models and strategies different from the traditional lecture-and-recitation mode, attempting to provide what Barbara Nielsen describes as "alternative active learning opportunities for the students." For Nielsen, some of the most effective methods involved role-playing exercises designed to take advantage of the *learning cycle* theory of instruction. These exercises—in which a specific chemistry lesson is broken down into the separate phases of exploration, term-introduction, and concept-application—offered students a way of experiencing a new concept, rather than relying on the usual textbook model of visualizing from a symbolic presentation, which even second-year students often find extremely difficult.

For example, Nielsen designed and videotaped a role-playing exercise to introduce the concept of dynamic equilibrium and its relationship to the behavior of strong and weak electrolytes in aqueous solutions, in which students act the parts of different species in solution:

> The activity is introduced as a game, and the students are given small cards, each with a colored star on it. . . .The colored species are allowed to mingle before the game starts. When the instructor gives the signal, [they] are to mingle further for an allotted time, [obeying specific rules appropriate to their chemical behavior]. After the allotted time all motion is stopped, data about the colored stars is collected and recorded, and a discussion about the data ensues. At this time, the terms of dissociation, dynamic equilibrium, amphiprotic species, weak and strong electrolytes, and weak and strong acids and bases can be developed. Further, the instructor can introduce the symbolism commonly used in textbooks to aid the students when reading the associated texts.

Students also write about the exercise in class journals, evaluating its effectiveness as a learning tool. Their entries have convinced Nielsen that the activity makes complex ideas more accessible as well as more memorable to them, and some students have noted effects that go beyond the immediate lesson, such as getting to know their classmates better.

Sandra K. Madison and James Gifford suggest that similar sorts of modeling activities might help to retain women and minority students whose frustration in their initial computer programming course discourages them from continuing in the field. Drawing on constructivist studies of how women and men learn, Madison and Gifford note that the usual emphasis on analytical learning styles leaves little room for students—many of them women—who employ concrete information-processing styles. In contrast, "providing concrete and graphical representations appears to help both groups of students understand the analytical programming concepts." In other words, like many of the pedagogies introduced through the Women and Science project, role-playing and modeling exercises may be designed to help a particular category of student, but they benefit many students.

John Krogman, a Faculty Fellow and Chair of General Engineering at UW-Platteville, also emphasizes the importance of being sensitive to different students' learning styles and experiences. This has often meant incorporating "teaching methods that include more student interaction and participation":

> Our Introduction to Engineering and Engineering Graphics courses, largely taken by new freshman, now both include a team project experience. This experience includes working together as a team of 3-4 students, doing some research or a simple design and then making both a written and oral presentation. It really has improved the interpersonal and communication skills of all of our students.

He has attempted similar innovations in his own Introduction to Engineering courses, by pairing students for computer assignments:

> Early in the semester, I will ask the students to categorize themselves as "experienced" or "inexperienced" with computers or the software we're using. I then . . . [try] to match an inexperienced student with an experienced one of the same gender. . . . I truly have found that "learning by doing" works very well, and the women initially seem to be more comfortable working with each other. Later in the semester, as the students become more comfortable with the software, I can randomly assign partners without any student afraid of being intimidated by someone with more experience. This overall approach . . . gets the entire class to the proficiency level needed much more quickly than earlier approaches.

Madison and Gifford note that classroom activities that use concrete simulations to communicate difficult concepts—such as having students role-play program modules or illustrating variable swapping by moving eggs or pieces of paper—give class members a vivid way of recalling specific computing processes. Analogies can have the same impact, as when an instructor likens memory variables to envelopes placed in different kinds of mailboxes. "Although . . . many students initially reject the lesson as juvenile in nature, . . . a majority of them confide later that it is highly effective."

## Restructuring Courses

While many participants designed fairly complex classroom exercises, others found that changes that seem considerably less elaborate can have far-reaching results. For Brian Bansenauer, the apparently simple act of rearranging the usual order in which topics were presented in his UW-Eau Claire algebra course made a significant difference in students' ability to understand basic concepts. Rather than introducing material in what he saw as the textbook's "disjointed presentation, [which] seemed based around the idea of mastering certain techniques . . . that the student would need later," he decided to introduce those techniques only as students actually needed to use them. Bansenauer created question sheets that "divided and organized the material, emphasizing concepts to understand rather than rules to memorize," and then based classroom work, quizzes, and exams on those questions. "This seemed to broaden the framework for understanding the material," which became focused not simply on how to solve particular problems, but also on questions about how those problems fit with others, what distinguished them from one another, "and what would happen if we extended the basic idea in another way." Students seemed to apply and retain information in a more coherent way and, writes Bansenauer, "I noticed a remarkable increase in the energy level students brought to learning methods to solve a problem when presented in context with the broader concept."

Heidi Fencl, the current Director of the UW Women and Science Program, also considers course organization fundamental, and her experience in revamping an introductory physics course for biology majors is exemplary. She did some basic restructuring of the course—for instance, focusing content more narrowly and adding student presentations on individual physicists in order to reinforce the fact that "science is a human endeavor." The innovations Fencl devised for courses at both a liberal arts college and a UW campus seemed to have such a powerful impact on students that her description is worth quoting at length:

> I found that the most important thing I could do for the peace of mind of my students was setting the framework for my expectations on the first day. Most of the students were sophomores or juniors [planning medical or health-related careers], and as such had predetermined ideas of what a science course should look like. From the beginning, I explained not only how this course might differ from their expectations, but also why the changes were appropriate for *them*. . . . Comparing medicine to physics led to a discussion of science as a process, and set the development of information gathering and problem solving skills as a higher objective than learning facts. I began each unit with a return to these priorities, and connected the upcoming sections with both past material and overall course goals. Once I began framing the course in this way, comments to the effect that "physics has no relevance for my life and major" virtually disappeared from my course evaluations and conversations with students. Discussion of course objectives, and how and why they were set, is one example of what is called metateaching. Throughout the course, I continually called attention to the learning process as well as to the material. [We] discussed how to study physics, how learning styles would manifest themselves in the material, and even how to deal with frustration. . . . While I believe that metateaching enhanced the atmosphere of working together towards a common goal that was so important to the way I taught the course, it needed to be backed up with some real changes in the way the material was approached. Friday Sessions, as they came to be called by my students, were the most visible change to the lec-

> ture section. Fridays were set aside so that the entire period was spent with students working on questions and assignments in a small group setting. There were informal, as well as formal, components to the way I structured these sections. . . . I did not assign students to groups or assign roles within groups. I also allowed groups to use the time according to their own needs, and to prioritize discussion topics in ways that were appropriate for them.
>
> These informal aspects to the group sessions made it important that the tasks assigned to the students be productive. Each Friday, I gave students their assignment for the next week. The assignment included reading, problems and conceptual questions. In addition to taking questions from the book, I wrote some of my own, and found both Arnold B. Arons' *A Guide to Introductory Physics Teaching*[3] and collaborative learning materials from the physics department of the University of Minnesota to be excellent resources for thought-provoking questions. Students did not turn in their assignments for a grade. Instead, the following Friday they were given points based on the percentage of problems and questions that they had attempted, and the group time was devoted to continued discussions of that assignment.
>
> The reduced emphasis on grades did not result in lower performance on the assignments, and, in fact, it allowed me to include more complex questions and problems than would otherwise have been perceived as fair. There were few students who chose not to participate collaboratively (once I determined that this was by choice, I did not force the issue); nearly the entire class was very active and enthusiastic in their discussions. The seating in the lecture room was not conducive to group interactions, and so we also used the laboratory for Friday Sessions. This accidentally introduced a wonderful feature: students would often use equipment from that week's laboratory exercise to design their own mini-experiments or to illustrate a point for others in their group. The camaraderie that grew between the students also set the stage for spirited discussions, and a visitor to one such session later commented that the students were actively involved with each other and seemed to truly value the importance of learning from each other.

Fencl's "Friday Sessions" represent a major change in the organization of her course. Other project participants found it possible to introduce student-centered methods in smaller doses. In his algebra course, for instance, UW-Eau Claire Faculty Fellow Alex Smith found that by orienting a workshop-format section (or math lab) around the discovery approach, he could encourage students "to make observations, make guesses, communicate their formative ideas to fellow students, learn from their mistakes, and eventually to build definitions and to discover concepts and methods of the course syllabus on their own. In short, in this setting students [could] be encouraged to process [as] mathematicians proceed."

Like many of the practices project members introduced into their classrooms, this workshop approach included dividing students into groups whose size depended on the assignment (for instance, two or three students for spreadsheet projects, four or five for work involving transparencies), and then having each group present their results to the class. "By not lecturing and instead constantly placing the responsibility on the *students* to build definitions and develop concepts, an instructor can hope that the student will begin to appreciate the attitude of critical thinking and . . . even make it a habit."

## DVPs and the Collaborative Community

A major component of the collaborative-community side of the Women and Science project involved having the DVPs model new pedagogical approaches for their colleagues in science, math, computer science, and engineering. By combining a number of different teaching practices in her introductory biology course (taken by both majors and non-majors), Distinguished Visiting Professor Judith Heady not only provided her UW-La Crosse students with a more active learning experience, but also demonstrated these methods for other faculty members:

> Instead of lecturing I prepared sets of questions for each assignment and "graded" these and [the students'] oral class participation . . . . Class periods consisted of answering questions I found from reading their answers on the question sets, working in small groups on problems from the question list or from other sources such as the news, student reporting from the small groups, working on projects [with] self-appointed leaders sharing [the results] with the rest of the class, and sometimes compiling on the board the high points of student written and oral answers.

Heady's lab sections were organized into three-week sequences on the course's major topics (the environment, genetics and evolution, chemistry and cells, and energy), with students planning and carrying out basic experiments, keeping lab notebooks, and preparing research reports in small groups. Throughout the term, they were encouraged to give Heady feedback and to compare their experiences to their own goals for the course. Student responses to this approach illustrate both the advantages of her student-centered classroom practices and the problems teachers encounter in employing these methods. She was disturbed, for instance, that an extremely promising woman student dropped the class because she "lacked self-confidence in her learning and felt that she could not get enough information with these classroom methods." This was not the only student concerned about covering material they thought would be needed in later courses (another one was "afraid other classes have passed us by and we have fallen behind"). Still, many others noted that they were "learning how to answer why instead of just what," were "looking at science in a more scientific manner now," and even remarked that "It's harder to understand but much more satisfying."

Nevertheless, the problem of teaching students what they need to learn, especially in introductory courses, is one that faculty must take seriously, even as they try to challenge themselves and their students to be more active collaborators. Heidi Fencl, for example, writes that "Helping students explore course material is what I consider to be the purpose of the laboratory section of an introductory physics course. While I don't downplay the importance of teaching laboratory skills, there are many opportunities for those lessons to take place. As a result, I replaced experiments designed only to teach techniques with exercises designed to help students explore content." Yet at the same time, she emphasizes the importance of remembering why students are taking a particular course, especially when it comes to evaluating their performances:

> I continue to wrestle . . . to find an effective method of testing. While a timed, in-class exam is clearly a stressful experience, most algebra-based physics students take the course at least in part as preparation for a standardized test such as the MCAT. I therefore felt some obligation to test them in the course as they would be tested on the material in the future. As a result, I used an in-class exam, and included a variety of question types, from multiple choice to essay. Stress of the examinations was reduced somewhat by [allowing] students to rework any question or problem for which they did not receive full credit on the exam. They could receive up to one-half of the points that they missed on any given question, but I required a greater degree of clarity and explanation on the regrade than I did on the original work. (For example, students were not allowed to merely choose another answer on the multiple choice questions; they needed to justify their new selection and explain why it was a better choice than their original answer.)
>
> I had some qualms the first time I tried this regrading procedure. I was concerned that the students would object to the additional work, and that the grades would be artificially high. I found instead that the students were much happier, and put a great deal of effort into their regrades. The exams, while still not popular, became more of a learning experience and less of a trial. Students had a motivation to continue to work with unclear material, so exams no longer closed the door on a unit as they had done previously. By comparing scores to previous classes in which I did not regrade exams, I felt comfortable that the students did not receive higher course grades than they earned. However, their comfort level with seeing an 85%, as opposed to a 77%, as a B was certainly higher, and this manifested itself in an improved attitude toward the course and a better work ethic.
>
> . . . As we discussed learning styles, I acknowledged that some students would catch on to physics quicker than others, and that my concern was that they learned the material and not when they learned it. In fact, some of the best physical understanding came from students who struggled early in the course. Students whose grades increased throughout the semester, and whose final examination showed an understanding of early material, were given a final grade that reflected that understanding. Again, that allowed me to write exams at the level I wanted my students to achieve, and to set my expectations high. It

> also encouraged students to continue to try, and to know that successful effort would be reflected on their transcripts.

Fencl remarks, too, on the difficulty of knowing the precise effectiveness of these changes in pedagogy and course structure, and the necessity of using both formal and informal measures to evaluate the usefulness of these student-centered approaches:

> I was not involved in any controlled experiment to compare either learning or satisfaction of students in courses with different pedagogical bases. However, anecdotal evidence suggests that cooperative pedagogies were successful for me in both ways. My students reported good performance on physical science sections of the MCAT and practice MCAT, and I saw a qualitative difference in student work as the course evolved.
>
> Student evaluations were also higher than the college average, which was surprising for a required, algebra-based physics course. Comments repeatedly referred to the usefulness of the group sessions and in-class experiments, and to the appreciation of the students for the variety of chances to ask questions. My favorite comments, however, came regularly from students who stated that they dreaded taking the course but ended up enjoying physics and finding the class to be beneficial.
>
> Another indication of the success of this approach is that my drop rate for the algebra-based physics course was almost zero, and the enrollment grew by almost a third over a three-year period. Informal surveys that I distributed also indicated that students in these classes involving a mixture of pedagogies had increased levels of confidence and satisfaction with the material and the experience. This was true for both men and women, and was perhaps best phrased on a student evaluation from several years ago stating "Frequently enjoyed lab with reckless abandon."

* * * * *

The preceding examples of pedagogical innovations have implicitly assumed that many aspects of science learning and teaching do not depend on the cultural backgrounds or identities of either student or teacher, but there are many ways in which cultural differences do play an important part. In this section, Catherine Middlecamp, a chemist at the UW-Madison, tackles the goal of expanding curricular reform and faculty development so that it becomes sensitive to an even wider array of underrepresented student populations. (Her suggestions for trouble-shooting appear at the end of the essay.)

# How Can We Improve Our Science Teaching? A Case for Cultural Knowledge

**Catherine Middlecamp**

Cultural mismatches between faculty and students easily can lead to misunderstandings and miscommunications. For example, if a traditional or bicultural Navajo student is asked to dissect a frog, the student may choose not to return to the biology laboratory.[4] If the lowered eyes of a quiet Hmong student are mistaken for lack of interest, the intellectual potential of this student may go untapped.[5] If the different language patterns of an African-American student go unrecognized, the student and teacher may fail to communicate and both may become discouraged.[6] Thus, those who can teach with sensitivity towards a variety of cultural norms for behavior are better equipped to facilitate the learning process than those who cannot. With such knowledge in place, teachers can build bridges that are respectful of both the culture of students and the culture of science.

Cultural mismatches also can lead to unconscious and consistent biases against particular groups of students.[7] For example, when native English speakers listen to those lacking fluency in English, they unknowingly may discourage those students' further participation in class activities by their facial expressions (e.g., a pained expression of intense concentration, with furrowed brow). They may never learn the names of students who have "strange" names. . . . Clearly, any such actions are problematic: Students who do not feel respected, connected, or safe are less likely to feel motivated to learn.[8] These issues are especially troublesome when a student's participation in class is subtly discouraged, yet linked to the course grade.

The case for acquiring cross-cultural knowledge also rests on the promise of better learning for *all* students. By including a variety of scientific examples relevant to different cultural groups, all students have a greater chance of connecting with something meaningful. By designing teaching practices to benefit minority students, majority students benefit as well.[9] Furthermore, when students are exposed to cross-cultural perspectives, they are better prepared for employment in the global community. They can meet the needs of a variety of people in health-related professions, in industry, in business, or in their home communities.

Our nation needs its *entire* pool of scientific talent. As Shirley Malcolm points out, diversity serves the national interest.[10] Science needs people to articulate the concerns of those whose voices have not been heard. Scientists need to appreciate and include the wisdom of other cultures, including traditional agricultural, environmental, and medical practices. Students of science need more role models to speed along the process of supporting new practitioners from underrepresented groups. Thus, the entire scientific community stands to benefit from including a more diverse group of practicing scientists who can contribute both to the practice and teaching of science.

Finally, a more diverse group of scientists should facilitate a more inquiring stance towards science itself. For example, a traditional Navajo perspective might better situate science in an ethical context, and help us pose questions such as, what knowledge is important to the surivival of our society, our earth? To what use will knowledge be put? What might be the effects of using knowledge in this way?[11] With the wisdom of more reflective cultures, it may become possible to more routinely critique science in a way that realistically assesses its strengths and weaknesses. Such a critique could empower us to envision different relationships between science and contemporary societies.[12]

## Teaching Science with Cross-Cultural Examples

Consider Maria Elisa, a student from Puerto Rico who is entering a state university in the Midwest. She is a native Spanish speaker, but has studied English since grade school. She leaves behind a tropical island where one can grow coffee and mangos, and where the ocean (and its hurricanes) is ever present. She now finds herself in the land of cornfields, smack in the path of summer tornados (and winter arctic blasts) and will find salt water only as puddles on winter roads.

Maria Elisa does not represent an isolated case; rather, she is one of hundreds from Puerto Rico who study at Midwestern universities. Will she and those like her feel welcome in the science classroom? In part, the answer depends on her peers, her own personal resources, and the overall "climate" at her university. Her instructor (who most likely will be white and non-Spanish-speaking) also will play a significant role.

It is not difficult for those of us who teach to reach out to students from Puerto Rico, or for that matter to students from any culture that differs from the one that prevails. It does, however, require some initiative on the part of the instructor. One way to begin is with the basics, such as with a map. For example, ten years ago, I framed and put up a tourist map of the island of Puerto Rico. Now, when I meet with students from Puerto Rico, we can chat about their city of origin. In the process, the students have taught me about the culture of the island. I went one step further and took a semester of Spanish. This modest investment of time has paid surprising dividends with both my students and my colleagues.

To better include the life experiences of Maria Elisa and my other students from tropical climates, I began collecting culturally relevant examples for teaching. Consider, for example, a person wearing glasses and working outside on a sub-zero day in Green Bay, Wisconsin. When this person comes indoors, her glasses will fog up. In Puerto Rico, however, the opposite is true. Upon *exiting* an air-conditioned building and stepping into the warm humid air, her glasses will fog. . . . By using more culturally diverse examples, . . . the scientific principles involved can be more fully explored to the benefit of all students.

Different cultural settings can lead one to ask different scientific questions. For example, a group of us that included a Puerto Rican visitor were discussing a project for the analysis of radon from sites across the state of Nebraska. Eventually the question popped up, "Is radon a problem in Puerto Rico?" At first, we speculated [about] whether radon was present in significant amounts in the island subsoils and rock. From our collective knowledge, we knew that in geologic time, the island was old and volcanic. It also contained substantial limestone deposits, as evidenced by the region of karsts and limestone caverns. Although volcanic rocks should be a source of radon, none of us had heard that any radon testing was being conducted in Puerto Rican homes.

Our conversation quickly shifted. The key issue did not seem to be whether or not radon was present, but rather if the presence of radon *mattered*. We were familiar with the standard procedure for testing for radon in the Midwest—that of placing a test canister in the basement for an extended period of time. If significant concentrations of radon indeed were detected, improving the basement ventilation often would take care of the problem. However, basements and tightly sealed homes are rare on a tropical island. Most Puerto Rican homes are built on concrete (with no basement) and have louvered windows to maximize the sea breezes. Thus, even if radon were present, its accumulation in living spaces was unlikely to be a problem. We later found out that radon would be expected in the silica-rich volcanic rocks (rhyolite) of Puerto Rico. This discussion taught us more, however, than simply a few new scientific facts. In the process of switching the context to a different culture, we raised different questions and more fully explored the issues.

The lesson for those of us who teach is straightforward: our examples need to be more expansive. They need to include issues situated in different climates, geographies, styles of living, human body types, etc. The discussion about radon in the Midwest could easily be enlarged and made inclusive of a variety of settings. A discussion about car engines could be expanded to include examples relevant to those who don't use cars or whose values conflict with an automobile-based society. Genetic examples based in eye color or hair color could examine other traits that are more relevant to population groups who show little variation in hair and eye color. Cross-cultural examples need to be a mainstay of our teaching.

## TROUBLE-SHOOTING: FREQUENTLY ASKED QUESTIONS

For some, a consideration of culture and culturally inclusive teaching is quite unfamiliar. Here are some questions that may arise:

• *There are dozens of cultures and subcultures. How can I expect to learn about all of them?* Teachers don't need to become anthropologists (and anthropologists aren't savvy in all cultures, either). Attend to the cultural differences that matter in the classroom, such as the use of silence, body language, or personal values. Be on the lookout for your own biases, positive or negative, such as towards people who speak with a certain accent or prefer a different degree of eye contact.

• *What about stereotyping?* This can be a problem. Terms such as "white," "Hispanic," and "black" can be useful, but they paint with broad brushstrokes. They do us a disservice in lumping extremely dissimilar people together. Individuals simply do not neatly fit into categories.

• *Am I expected to create the perfect cultural match for each group of students?* No, and this is not the point. Rather, use your cultural knowledge to become more attentive to when an individual or a group is experiencing some difficulty in learning. Are your actions as a teacher contributing? Are printed materials contributing, such as by their biases in artwork? Are clashes between the school culture and the home culture contributing? Do some detective work and adjust things as you are able.

• *I'm a busy person. What is one step I might take?* Do a mental check of your "comfort level" with different types of students. With which group or groups do you feel the least comfortable? Look for an opportunity to follow up on this in your own way: reading, Web searching, attending a lecture, talking with others. If your own way includes doing a lot of talking, to learn more quickly you may need to improve your listening skills.

• *I give lectures. Any suggestions?* One is to attend to your use of questions in the lecture hall. How long do you wait for an answer? Who gets called on? Another is to check the examples you present; i.e., are you biased towards particular types (such as problems without much context)? Check your guest speaker list. Have you included some folks different from yourself? Ask your students for feedback: Is there something that you might better attend to so as to increase their comfort level? In pursuing questions such as these, you are looking for teaching practices that impact differently on different groups of students.

• *What if the needs of different students conflict?* This happens. For example, you may have students who are comfortable with cooperative work and others who prefer competitive work. You may have some students who answer very quickly and others who prefer to let more time elapse. Rather than viewing the needs as conflicting, you may wish to see them as complementary; that is, both are ways of acting needed for success and both should be part of the classroom dynamics.

• *Cultural tendencies such as preferring cooperation over competition won't prepare students for the real world. Doesn't avoiding competition do our students a disservice?* Actually, teamwork and cooperation are necessary skills to bring to the job market. A student versed only in competition will be at a disadvantage when needing other skills on the job.

## NOTES

1. Kenneth Bruffee, "Science in a Postmodern World," *Change* (September/October 1992), p. 20. Bruffee, a professor of English at Brooklyn College, CUNY, is also the author of *Collaborative Learning: Higher Education, Interdependence and the Authority of Knowledge* (Baltimore: Johns Hopkins University Press, 1993), and his work has been used as a guiding text at a 1996 series of Beloit College summer workshops for biologists.

2. Sheila Tobias, *They're Not Dumb, They're Different: Stalking the Second Tier* (Tucson: Research Corporation, 1990), p. 13.

3. New York: John Wiley and Sons, 1990.

4. Catherine H. Middlecamp and Omie Baldwin, "The Native American Indian Student in the Science Classroom: Cultural Clash or Match?" in *Proceedings of the Third International History, Philosophy, and Science Teaching Conference*, University of Minnesota, October 29-November 2, 1995, p. 783; Sandy Greer, "Science: 'It's not just a white man's thing,'" *Winds of Change* vol. 7 no. 2 (1992), p. 14.

5. Susan Katz Miller, "Asian-Americans Bump Against Glass Ceilings," *Science* vol. 258 no. 5085 (1992), p. 1224.

6. Eleanor Wilson Orr, *Twice as Less* (New York: W.W. Norton, 1987), p. 201.

7. Ray Lou, "Teaching All Students Equally," in *Teaching from a Multicultural Perspective, Survival Skills for Scholars*, Vol. 12 (Thousand Oaks, CA: SAGE Publications, 1994), p. 35.

8. Raymond J. Wlodkowski and Margery B. Ginsberg, *Diversity and Motivation: Culturally Responsive Teaching* (San Francisco: Jossey-Bass Publishers, 1995), p. 2.

9. Shirley M. Malcolm, M. Aldrich, P.W. Hall, P. Boulware, and V. Stern, *Equity and Excellence: Compatible Goals* (Washington, D.C.: American Association for the Advancement of Science, 1984).

10. Shirley M. Malcolm, "Science and Diversity: A Compelling National Interest," *Science* v. 271 (1996), p. 1817.

11. Sharon Nelson-Barber and Elise Trumbull Estrin, "Bringing Native American Perspectives to Mathematics and Science Teaching," *Theory into Practice* vol. 34 no. 3 (Summer 1995), p. 176.

12. Sandra Harding, *The "Racial" Economy of Science* (Bloomington: Indiana University Press, 1995).

*Chapter 4*

# Transforming Classroom Climate and Changing Course Content

**Laura Stempel with Cheryl Ney and Jacqueline Ross**

> Many faculty do not incorporate the work of women scientists into the curriculum, although an occasional woman of exceptional talent may be mentioned in a science class. Some textbooks and class materials may be overtly sexist; one still hears, for example, of science classes in which a slide of a naked or nearly naked woman is interspersed in a presentation, ostensibly to liven up the class.
>
> —Bernice Resnick Sandler, Lisa A. Silverberg, and Robert M. Hall, *The Chilly Classroom Climate*[1]

Chapter 3 outlined a variety of pedagogical innovations used by participants in the Women and Science project as they worked to attract and retain students who might otherwise abandon the science curriculum before their general education requirements were even fulfilled—despite having entered college intending to major in science or math—and those "second tier" students who are capable of becoming majors but are put off or intimidated by the curriculum and culture of collegiate science. As crucial as these practices are, however, two other classroom factors have an equally important impact on student attitudes toward and performance in the sciences: the specific climate in which teaching and learning take place and the content of the courses themselves. A landmark study by Seymour and Hewit explains why this is of particular importance for women students:

> We posit that entry to freshman science, mathematics or engineering suddenly makes explicit, and then heightens, what is actually a long-standing divergence in the socialization experiences of young men and women. The divergence in self-perceptions, attitudes, life and career goals, customary ways of learning, and of responding to problems, which has been built up along gender lines throughout childhood and adolescence, is suddenly brought into focus. [2]

From the opening days of the Women and Science project, participants have made it clear that the climate in which they teach and do research is one of their central concerns, and that is a concern of this chapter. But it will become evident that climate is a crucial issue for students as well, and questions about both climate and course content and curriculum are intimately connected and even fundamental to pedagogical concerns. Charles W. Schelin, then a dean at UW-La Crosse (now Provost at UW-Superior), cites an important study of freshman science sequences that attempted to discover why students who had previously intended to major in the sciences changed their minds, and how this affected women in particular:

> Poor teaching, faculty unapproachability, and the fast curriculum pace were the most frequently identified problems of both major "switchers" and "non-switchers" in [the Seymour] study. Both groups of students felt that the large classes, coupled with poor teaching and lack of faculty help, constituted an "unofficially legimitated weed-out system." . . . . The factors which had more significance for women . . . included poor teaching, discouragement from low grades in freshman and sophomore courses, and rejection of the lifestyle implied by particular [science, math, and engineering] careers. Many of these women found it difficult to learn from faculty who took no personal interest in their achievement. Often, they had progressed to this level in the pipeline with the encouragement of K-12 teachers who provided support for their talents and accomplishments. As a stark contrast, the university freshman science experience was cold and impersonal.

The task of creating a supportive learning climate for women and members of other underrepresented groups has been central to many curriculum-reform projects, including those institutionalized through Women's Studies, Afro-American Studies, and other area studies programs. There are at least three distinct levels of learning climate with which faculty must deal: a general, often campus-wide one, in which women and others may feel excluded from study or even denigrated for their interest in the subject; the climate within a particular major, program, or department; and the specific learning climate of the science classroom, in which teaching style and content are often based on certain assumptions about what sort of student is likely to go into and succeed in the sciences. Common stereotypes of such students too often paint a picture of an introverted young white male, math wiz, highly focused and motivated to do science, the "cream of the crop" and certainly not holding down a part-time job! As Cheryl Ney made clear in Chapter 2, revising those assumptions can be a key to encouraging more and better students to major in the sciences, and even to pursue graduate and professional careers, and these changes go beyond pedagogical technique. In a 1995 article on teaching college science, Teresa Arámbula-Greenfield indicates how deeply rooted problems of climate are:

> Girls and ethnic minorities have consistently scored and continue to score lower than do white male students in science, both on the whole and by level of profiency. . . . This is not surprising, considering that girls and boys experience different levels of participation and support within typical science classrooms as boys dominate teacher time, class discussions, hands-on activities

> and equipment, and so forth. . . . It is not only females who ultimately can lose out in college science programs, however. It has been demonstrated that many college science courses rely on outdated content and traditional pedagogies . . . [which] can be especially critical for female or minority students who may not function optimally in the competitive, expository teaching mode, which tends to characterize both major and non major science classes. . . .[3]

Before teachers can create a comfortable, supportive, and effective learning climate for the students, of course, they must themselves feel free to extend their own range, both in pedagogical practices and in course content. That freedom is one of the fundamental requirements for curricular change, and as later chapters will demonstrate in more detail, participants in the Women and Science project faced a wide range of response to the idea of change—from enthusiasm to hostility—from individual departments and administrators throughout the University: In some cases, faculty were committed to exploring new approaches and supportive administrators took an active role in the project, while in others, resistance among departmental colleagues made it extremely difficult to introduce pedagogical innovations, to design new courses with major components on gender, or to gather faculty for related workshops and lectures.

For example, as one of the project's National Science Foundation (NSF) evaluators reported after a 1994 site visit to UW-River Falls,

> The chemistry department, host to . . . DVP Cheryl Ney, was battling over revisions in their core curriculum, much of the battle concerning new classroom approaches, new hands-on labs, new approaches to the large lecture section. As we were leaving, several of the chemistry faculty—all involved in the Women and Science Program—whispered to us, "We won; important changes in the chemistry curriculum were approved by the department faculty over the vehement disapproval of a few."[4]

We will return to this issue in Chapter 5, when we look more closely at the Collaborative Community of which UW-River Falls was a part. For the moment, though, this variety in response is an important reminder of the role departmental and institutional climate plays for faculty, especially those without tenure, who must sometimes attempt to revise their courses in the face of collegial hostility and occasionally at considerable professional risk to themselves.

On campuses where there is already a supportive environment, change is obviously much easier. Sometimes this positive climate results from the fact that what is new to one person may already be familiar to another. At other times, it involves the recognition that what looks like change is actually the implementation of familiar techniques, as mathematician Alex Smith notes:

> To untenured faculty involved in the Women and Science Program [at UW-Eau Claire], . . . a climate was present in which they felt secure in experimenting with the use of cooperative learning groups. The existence of the climate made its presence known through department discussions organized by [DVP] Sherrie Nicol. In these discussions, it was found that faculty readily accepted the idea that there are in fact different learning styles and that there was benefit to teaching in ways that allowed students with different learning styles to thrive. In addition, during these discussions it became clear that tenured faculty who primarily teach courses for Math Education majors already used cooperative learning groups extensively and that they would enthusiastically support untenured faculty who wanted to experiment with such teaching methods.

Smith describes a situation in which untenured faculty were actively encouraged to take part in the program. "In part," he notes, "this was due to a vision that the most important reforms are brought about by grassroots efforts and that untenured faculty are the very roots of the future." His department is unusual in its interdisciplinary makeup—three of the math faculty members are actually math *education* specialists, who at other institutions would probably be found in an education department—but Smith's comments remind us of the role departmental and institutional support can play in faculty members' willingness and ability to be innovative. We'll return to this subject in Chapter 6, which examines in institutional terms the question of how systemic change is accomplished and made permanent beyond the classroom.

## A First-Year Experience

Danielle Bernstein's account of her own experience as a Distinguished Visiting Professor in the computer sciences at UW-Stevens Point provides a good outline of the basic problems confronting faculty involved in the project. She is also a role model for her path-breaking work on changing both the content and learning climate of the freshman or first-year course by making an introduction to the culture of the major and discipline itself part of the course's content. In fact, what she did in her teaching as a DVP can be understood as embedding what has come to be called a "freshman year experience" into the course.

According to Bernstein, researchers have offered a variety of explanations for the noticeable drop-off in women computing majors—20%, down from a 1986 high of nearly 39%, according to the National Center for Education Statistics. Many of these explanations involve men's and women's different attitudes toward computers and the great disparity in expertise with which students enter their first computing courses. Faced with these statistics, Bernstein and three Faculty Fellows surveyed their first-term students, asking them to rate themselves on comfort and knowledge with software and hardware. Through this survey and through faculty workshops at other UW campuses, Bernstein and her colleagues discovered that an introductory course that claimed to have no prerequisites actually had "hidden prerequisites" that inadvertently tended to favor students, most of them male, with greater computing experience.

For example, she reports, it turned out that computer science faculty assumed students would learn necessary jargon by hanging around with other students, rather than through classroom exercises—an idea which in turn assumed that computing students hang out together, participate in newsgroups, and use computers for fun outside of class. "While I was at UW-Stevens Point," writes Bernstein,

> a new course, "Information Tools," was added to the CIS curriculum. In the new curriculum, this will now be the first computing course for majors, before they are introduced to programming design. The purpose of this course is to attempt to level out the playing field between entering students. . . . Though computing curriculum issues had been discussed at UW-Stevens Point for two years, they were settled while the department participated in the Women and Science Program. At the time, everyone's sensitivity was heightened by the great disparity of knowledge between students coming into the department and by the hidden prerequisite of our current first Computing course.

The lack of information Bernstein describes presents an obvious barrier to certain students, and providing the basic tools in a field like computing is clearly crucial to creating a classroom learning climate that makes the subject accessible to all students. But Bernstein makes a key point when she adds that "the problem of the disparity of knowledge also includes keeping the advanced student interested and challenged. Several male students told me that they 'had been computing all their lives.' They hoped to see something different, now that they were in college."

Still, with the focus on the students who feel alienated from the sciences, Bernstein underlines the intimidating power of specific organizational features beyond the classroom. She notes, for instance, that "computer centers . . . operate as gatekeepers. They decide how users will interact with the central computer system" through the issuing of id's and passwords, and the imposition of rules about equipment access and operation, and that gate-keeping can result in the exclusion of students who find the process slow or the rules obscure. "All organizations have these problems," notes Bernstein, "but some students lose patience, especially if this is their first year in college. Their perceptions are that everyone else is on the system but them. They regard dealing with the computer system as another piece of red tape. But unlike registration and paying tuition, this particular piece of red tape can be avoided by dropping out of computing courses and not using the computer system." Rather than assuming, then, that everyone is equally prepared and motivated, Bernstein and her UW-Stevens Point colleagues developed a set of practices meant to insure that both beginning and advanced students will stick with the course until the network, equipment, and software become familiar. Among their suggestions for changing the classroom climate for learning, as well as for more successfully introducing students to the culture of the major and the discipline, are the following:

- Assume that students know nothing coming into the course.
- Survey the class on their perceived knowledge of various computing subjects.
- Share the results of the survey, as soon as possible at the beginning of the term. This might allay fears by some students that everyone knows more than they do. Surveying also allowed us to approach the undecided majors on the virtues of becoming a computing major. Undecided students are our opportunities to increase the number of majors.
- Explicitly teach how to get onto the Internet, use electronic mail and the World Wide Web. Do not assume that students will learn through the grapevine.
- Don't forget the needs of the advanced students. Introduce them to each other. Suggest more challenging work that they can explore.
- Discuss the time-consuming aspect of computing with students. Acknowledge that they have chosen a course and a major that take a lot of time.
- In class, discuss problems with the computer center. . . [but] try to encourage assertiveness. . . . When is it the students' problem and when is it the computer center's problem? And how can they tell the difference?

This list of suggestions is explicitly designed for computing majors taking their first college course. The point Bernstein makes about the time-consuming nature of course assignments, for example, is a particular issue in computer science classes, where the projects tend to be few in number but often quite lengthy, whereas most other science courses include a large number of relatively short assignments. Nevertheless, the underlying principles at work in Bernstein's recommendations clearly apply to a wide variety of science, math, and engineering courses, particular those with lab components. Equally important, as Bernstein indicates, "students need to belong to a community within their major," and one that goes beyond their mere simultaneous presence in the classroom. Group work, closed and supervised labs, and involvement in out-of-class activities help to facilitate a sense of community as well as providing students with positive experiences that encourage them to feel like professionals in their field. And science educators can look to Women's Studies, a field in which pedagogical issues like these have played a central part since its founding, to learn how to carry out such changes.

Bernstein adds some other specifics about how to encourage students from the underrepresented groups targeted by projects like this one:

> The very top students are usually recommended for research projects. But the vast majority of students do not go on to do research. Most computing professionals are practitioners. Ideally, there should be an opportunity for every student to get involved in an out-of-class activity in the major, beyond just being a member of a computing club. . . the faculty member should identify women students for positions as computing lab assistants, internships, and co-op assignments.
>
> At UW-Stevens Point, I gave a computing workshop in a career awareness day which brings high school girls onto campus. I asked two first-year CIS students to help me in the computing workshop and to act as role models for the high school girls. They were very surprised and pleased to be asked. They spent time and effort preparing their presentations and deciding [how] to present the right image. Their self-confidence in the course rose noticeably after that experience.

Once again, while this example deals specifically with computing majors, the lesson applies to students in a wide variety of science fields. By offering opportunities for out-of-class activities to people who do not fit the conventional notion of the future researcher, teachers may encourage a wider range of students to see themselves as real or potential science professionals.

Similarly, Sandra Madison and James Gifford, who were Faculty Fellows during Bernstein's tenure as DVP and who also teach computing at UW-Stevens Point, point to an insight whose implications go far beyond their specific field of expertise. Noting that research demonstrates that "programming is the computing component that is most inimical to females, the area where they are least likely to persist," Madison and Gifford offer an obvious solution: "one way to attract females to information systems and other computing-related fields is to improve their initial programming experiences." Once again, the same approach to improving the climate in which students learn should work in any science discipline—or indeed, in any discipline at all. By making certain that introductory courses are inclusive and that students are not presumed already to know what is needed for further work in the field but are instead given a comprehensive understanding of basic concepts, faculty members make it easier for students outside the traditional pool to imagine themselves as science majors, graduate students, and professionals.

## Introducing Gender, Race, and Class

The second half of this chapter discusses the importance of modifying the content of science courses to include the introduction of issues related to gender and science. For some students, however, such changes, along with pedagogical shifts such as those discussed in Chapter 3, can produce a specific climate problem of their own. Distiinguished Visiting Professor Judith Heady taught a Senior/Honors

Seminar on "Gender, Race and Science," and discovered that students were not always comfortable discussing these topics in the classroom. Here again, students' experiences in previous courses played a role, although in this case it was experience in courses outside of the sciences:

> After some introduction and struggling to get the 17 women and 8 men to discuss topics as a whole class and somewhat more successfully in small groups, [I had them do] oral presentations singly or in pairs on topics ranging from historical looks at a woman in science, women and men in sports and the military, feminist science, nursing, gender differences in spatial perceptions, hormones, [the possibility of] a homosexual gene, and other aspects of gender and race. Some women in the class voiced the feeling that they did not want to upset men by discussing the topic of discrimination [against] women in science. Other women with some Women's Studies backgrounds were more comfortable discussing historical and current problems.

In Heady's experience, then, a familiarity with feminist ideas made certain students more willing to engage with timely and controversial topics involving gender and science. Whether these students were more comfortable questioning traditional ideas about scientific research, accustomed to a classroom climate in which people disagreed with each other, or simply more familiar with the study of gender, their Women's Studies experience apparently prepared them to engage the social issues Heady wanted to introduce.

As Heady's experience illustrates, it cannot be assumed that students bring to their science study the kind of critical background that enables them to engage the kinds of topics she describes. Instead, a fundamental part of the Women and Science project agenda was changing the science curriculum in order to include discussions of cultural context that prepare students to consider questions about gender, race, and class. Some participants in the Women and Science project came up with specific classroom projects designed to confront students' preconceptions about the sciences. One especially self-reflexive activity that grew directly out of the project provided what Sandra Madison and James Gifford describe as "an instructive example of how such concrete exercises can also be used to foster a climate of inclusiveness in the classroom." Students on mixed-sex teams were told to prepare instructions for reassembling a Tinker Toy model as quickly as possible. The models consisted of a set of easily duplicated subassemblies, so that the most efficient plan would involve students building several sections at once, assigning different portions to individual group members. After writing the instructions, students discussed the exercise in class, but even more important, they answered a questionnaire about how their groups worked:

> The questions were designed to probe the unconscious bias of the students by exploring how gender influenced the roles individual students played during their work. Almost without exception, women were assigned the job of secretary or recorder, [and] one or two males assumed the leadership position, mediating conflict and assigning tasks, and presenting the final results to the instructor. A class-long discussion of why such gender-based divisions took place opened the eyes of many students of both genders to the unconscious stereotyping that they were part of.

Faculty members committed to changing the way the sciences are taught must find ways of overcoming what students have already learned about gender, as well as about science—an especially complex task, since, as this exercise demonstrates, students are not always aware of their own assumptions. While many classes begin with a consideration of the scientific method, few faculty use this opportunity to talk about how scientists practice science. Madison and Gifford's examples illustrate how issues of climate can be brought directly into classroom discussion and made part of the course content.

* * * * *

In the rest of this chapter, plant pathologist Caitilyn Allen outlines a variety of changes in content designed to make science courses more appealing to the groups of students that projects like this one are intended to attract and retain, particularly through the introduction of issues concerning gender. While some of her examples necessarily focus on her own specialty, the lessons Allen offers are easily applied to other fields, and additional examples from other areas of study appear in the Appendix. Faculty members are often intimidated by the thought of revamping the content of their courses because they imagine themselves having to learn vast quantities of new material—a daunting prospect, indeed! Allen's examples, which range from the inclusion of simple facts to the development of complicated classroom exercises, make it clear that considerable effects can actually be accomplished with relatively small changes in content. Often, in fact, teachers need only remember to incorporate into their lectures and readings information and ideas they already possess, and, as many project participants have already noted, to make sure that their presentation of the scientific and technical professions accurately reflects their own experiences and that of others they know.

# Shades of Gray: Changing the Content of Science Courses to Include and Encourage the Underrepresented

Caitilyn Allen

## Introduction

As it has become increasingly obvious that traditional approaches to science teaching have effectively excluded many women and minority members, science educators have begun to recognize the necessity of changing our habitual teaching patterns. However, to date most efforts have focussed on developing improved pedagogical approaches and methods that will make science classes more inclusive.[5] Examples of such new teaching techniques include co-operative learning and student-generated hypothesis testing. Innovations such as those described in Chapter 3 unquestionably represent an important way to broaden the appeal and effectiveness of our science courses. They bring the excitement of scientific discovery to undergraduate courses and give students from traditionally underrepresented groups a chance to visualize themselves as researchers, discovering something new about the natural world.

But in addition to how we teach, we must also consider reforming some aspects of what we teach. Science taught as a collection of facts is uniquely unappealing to many women and minority students: this approach generates an image of a black-and-white world where every question has only one right and many wrong answers.[6] Portraying science as an analytical activity exploring shades of gray—which, as every working scientist knows, it certainly is—makes it much more appealing. If we explicitly describe the practice of science as a fallible and complex human pursuit, we allow students to see scientists as mere mortals and thus to imagine themselves as scientists.

Further, most of us are probably trying to teach inappropriate quantities of information in our classes, especially in introductory science courses where students are making the crucial decision about whether to persist along a science path or change majors. A convincing set of empirical research results suggest that the habitual obsession with covering large amounts of factual material in science courses actually decreases students' comprehension of course material and generates a negative attitude towards the sciences.[7]

In practice, changing the content of science courses to make them more appealing and accessible to underrepresented students is not only a question of revising what is usually included, but also of considering what is commonly left out. Traditional undergraduate science courses, particularly at the introductory level, rarely address such issues as the complexities and ambiguities inherent in data analysis; the social context in which research is conducted both within and outside the scientific community; and the ethical, moral, and philosophical questions that arise concerning various aspects of both scientific research and its ultimate applications. Since these interesting shades of gray are unquestionably a part of the real scientific world, stripped-down factual science classes are thus not only dull for many students, but also in a sense unrepresentative. In this section I will present five suggestions for a more humanistic and multidimensional approach to teaching science, together with examples of each approach.

## METHODS

1. *Mention work by women and minority scientists in lecture and reading material.* This is the first and easiest step; in many cases, it is simply a question of scrupulously including first names. For example, in describing the famous Hershey-Chase Experiment that in 1952 established with beautiful simplicity that DNA and not protein is the genetic material, an instructor who called it the Alfred Hershey and Martha Chase Experiment would teach with no additional comment

that women have made important scientific contributions. In this experiment, viruses that infect bacteria were grown in the presence of two different radioisotopes, one that labeled the protein and the other that labeled the DNA. Hershey and Chase found that after infection, the viral DNA was inside the bacteria but the viral protein was still outside and therefore could not be the virus's genetic material. The key technological breakthrough in the Hershey-Chase Experiment was the use of a Waring blender to separate infected bacteria from the empty virus coats, a practical suggestion from a female colleague, Margaret McDonald.[8] Including information at this level of anecdotal specificity in lectures increases student retention and creates a more vivid and realistic image of the research process.

2. *Discuss effects of science on issues of interest to women and minorities.* Often students report feeling alienated from and uninterested in science classes because they perceive that science and technology have no effect on the lives of people they care about. It has been suggested that explaining the social relevance and application of science creates a powerful incentive for students in underrepresented groups to learn scientific material. For example, a lecture on the effects of the Green Revolution in agriculture ordinarily discusses the impressive increases in grain yields that were obtained when improved varieties of wheat and rice were planted in India and Mexico in the 1960's. A thoughtful lecturer might add that because these improved varieties demanded large inputs of fertilizer and pesticides, their widespread planting changed traditional agricultural practices. They required expensive fossil-fueled machinery, and occasionally caused environmental degradation. But in addition, the conversion of small family farms into the large acreages necessary for efficient mechanized grain production moved agriculture from the barter economy, traditionally the province of women, to the cash economy, a male preserve in traditional societies. Thus, another unanticipated outcome of the Green Revolution was a substantial economic disempowerment of women. Although I once mentioned this last effect in a minor aside (perhaps 45 seconds) during a 50-minute lecture on the Green Revolution, I was surprised to discover that the idea was mentioned in nearly half the student responses to a general essay exam question about agricultural reforms. Eighty-eight percent of the students who had retained that point were women, suggesting that personal interest creates a substantial motivation to understand and retain scientific material.

3. *Present science as a social activity.* Among the several disservices done to science by popular culture, the worst may be the creation of the stereotype of the scientist as a socially inept eccentric who works all alone in his laboratory. It's no great wonder that highly socialized students choose not to pursue a field of study they expect to make them lonely (and peculiar). However, in reality, scientific research is often extremely interactive and social. This stereotype can be reversed by teaching students some history of science as various concepts are explained, emphasizing the historical power of collaborations. For example, the lifelong col-

laboration between Otto Hahn (a chemist) and Lise Meitner (a physicist) resulted in the Nobel Prize-winning proof that radioisotopes decay into other more stable elements when they emit subatomic particles, and that atomic fission was likely to liberate huge amounts of energy. Letters between Meitner and Hahn demonstrate the intensity, power, and intellectual intimacy of their working relationship; it is unlikely that either would have been able to design and conduct the critical experiments alone. Teachers can point out that even highly individualistic researchers depend heavily on the scientific efforts of their colleagues and predecessors. (It should be pointed out, however, that unjustly, Hahn but not Meitner was awarded the Nobel Prize for this discovery. Based in large part on the record provided by the correspondence between Hahn and Meitner, recent discussions of this discovery acknowledge that Meitner was inappropriately overlooked by the Nobel Committee.)

In the same vein, different disciplines within the sciences have often enriched each other. For example, scientists who studied plant diseases were for many years puzzled by a group of apparently viral diseases that were inexplicably cured by antibiotics and other bactericides, although no bacteria were ever isolated from infected plants. One day a Japanese plant pathologist brought home an electron micrograph he'd taken of the perplexing infected plants and was studying it at the kitchen table. His brother-in-law, a veterinarian, glanced at the picture and said, "Oh, so plants get mycoplasmas too?" Thus the mystery was solved; the diseases were caused by odd-looking group of bacteria called mycoplasmas, which are obligate parasites and therefore cannot grow on any culture medium. They had long been known and studied as animal pathogens, but the perhaps excessively specialized plant pathologists had been unaware of their existence.

4. *Present scientific research in its social and historical context.* Based on what they learn from many introductory science courses, students might understandably imagine scientific research taking place in a social and political vacuum, timeless and hermetically sealed off from the flow of history. This is demonstrably (and interestingly) untrue. Although scientists are fond of seeing themselves as objective searchers after the truth, most of us recognize that we bring a set of biases, expectations, and perspectives with us when we enter the laboratory. These preconceptions can alter the hypotheses we pose, the methods by which we test our hypotheses, and even the way we interpret our results. As a consequence, scientific research is often an intensely human activity, colored by the social context of the moment in which it is conducted. At a comfortable historical distance, we can find many examples of profoundly biased science ranging from the embarrassing to the shameful. This is especially true of biology relating to humans,[9] but social expectations have also influenced research on chemistry, physics, and astronomy. This is not to suggest that there are no knowable facts; unarguably there are, or there would be no technology and indeed no particular point to most scientific labor.

However, students need to realize that although we can learn certain individual facts, the larger truths we construct from them are never wholly objective.

Paradoxically, most scientists agree that this degree of uncertainty makes science more rather than less appealing. Therefore, it is useful for teachers to offer students historical examples of bias in science and explain that current work is probably biased by existing social expectations in ways we cannot perceive or perceive only with effort or help from people outside our own discipline. Thus, the entry of researchers from previously unrepresented groups into a scientific field often improves the quality of the science in the area. There is probably no better example of this observer effect than the recent revolution in the field of primate social behavior caused by the entry of women scientists into this field.

Until the mid-1970's, the general dogma in social primatology was that troops of female primates (referred to by scientists as "harems") were dominated by a single powerful male who enjoyed exclusive sexual access to them. This male would protect "his" troop and fight off challenges by other aspiring dominant males. Female primates were seen as passive and submissive, a resource to be conquered. However, when female primatologists like Jane Goodall, Dian Fossey, and Sarah Hrdy began to study social behavior, their identification with female primates allowed them to perceive the same interactions differently and more accurately. It became clear that different primate species have highly diverse social structures and that behaviors vary depending on environment, even within a single species. But most interestingly, it emerged that in general, primate societies are matriarchal, with females forming the core of each group, while males are more loosely attached. Far from being dominated by one male, many female primates are sexually promiscuous, often strategically so. Earlier male primatologists were misled about these interactions because they focused their field observations on the individual identified as the dominant male. They did not observe the female-female interactions that would have allowed them to understand the troop's matriarchal power structure, nor did they notice that females often copulated with males other than the "dominant male." This striking paradigm shift has been attributed to the fresh and fundamentally different perspective that informs female primatologists doing field observations. Including this kind of information in lectures and readings effectively demonstrates both the difficulty of objectivity and how lapses are eventually corrected by the scientific process at work. Further, it also teaches by implication that diversity of approach and perspective is methodologically powerful, and that each individual has the potential to improve or even reconstruct a part of science.[10]

5. *Consider the ethical aspects of scientific problems, both historical and current.* In addition to studying the effects of society on science, the effects of science on society must also be considered. This aspect of science has long been studied by

sociologists of science and ethical philosophers, but it is usually not considered proper territory for working bench scientists—and certainly not for undergraduate students. However, good arguments can be made for bringing ethics into the science classroom. First, ethical dilemmas usually have a strong human-interest element that makes them compelling and memorable to undergraduate students. Second, asking students to engage in an ethical discussion forces them to use higher-order thinking skills, integrating different types of argument.

For example, a discussion about genetic engineering might consider the following two questions: Is genetic engineering of living organisms potentially dangerous? and, Do humans have the right to alter the genes and characteristics of living things? The first question requires students to imagine specific biological worst cases and evaluate their likelihood and possible consequences—essentially, a scientific line of argument. The second question, however, is moral or ethical in nature and demands an entirely different form of argument. An answer constructed to address the first question is meaningless in response to the second one. Confronting this truth demonstrates for students the limits of a perspective that is exclusively either scientific or moral.

Finally, many undergraduates have strong but largely unexamined opinions about ethical matters. Studying a morally and scientifically complex problem forces students to either modify their opinions as they perceive a more complex and nuanced world or, if they choose to retain their original viewpoint, to develop reasonable arguments in support of it.

A useful approach to generating a productive ethical discussion in the science classroom is to set up a problem by giving students enough factual background information so that they understand the context of the problem. Ideally, it should be directly connected to concepts covered in class. For example, a genetics class might consider whether researchers should attempt to change the human germline to cure a specific inherited disease—let us say, sickle-cell anemia. Different groups of students can be assigned specific points of view and allowed to do some research outside of class in order to formulate their assigned viewpoint's arguments. Each group is then asked to present their perspective in class in a structured format—perhaps as testimony before an imaginary Congressional subcommittee. Because it is understood that this is a kind of intellectual game, the student is relieved of the anxiety-provoking burden of exposing and defending his or her own personal opinion on the subject.

In this kind of mock hearing, a biologist might argue that the results of such a procedure are predictable and relatively safe, since the molecular basis of the disease is completely understood and the technology for germline change is well developed in animal systems. A theologian could respond that this genetic manipulation would represent a usurping of divine creative power to which humans are not entitled. A mother who had lost an 11-year-old daughter to sickle-cell anemia might plead on personal and emotional grounds for scientists to prevent any more tragedies like the one she experienced. A lawyer could point out that although this particular experiment might appear innocuous and humanitarian, it sets a precedent that could allow repellent or frivolous manipulation of the human genome in the future. Additional viewpoints might include that of the biotechnology company's insurance agent, an ethical philosopher, a representative of a federation of HMOs, the chair of the Senate Appropriations Committee, and so on. At the end of such an exercise, students could be asked to write a brief essay on the development of their own perspective on the issue, which would be graded as evidence of individual work.

The challenge of designing an exercise of this kind is to make credible and relevant as many different perspectives on a given scientific controversy as possible. If the exercise is successful, students come away from it with a vivid sense of science at the intersection of complex and conflicting forces, as well as a thorough understanding of the particular scientific concept underlying the issue in question. To successfully complete the exercise described above, students must learn a great deal of fundamental scientific material. They will learn how the normal hemoglobin protein functions, how a particular molecular defect results in sickle-cell disease, how genes are cloned, how transgenic animals are generated, and how red blood cells are formed in the stem cells of the long bone marrow. But in addition, they will have learned how to apply biological information to practical real-world problems, and will have gone beyond memorization to integrated big-picture thinking. It is this last skill that is ultimately the most important and enduring intellectual tool a student can acquire.

## Conclusions

Critics have suggested that changing the content of undergraduate science courses is virtually impossible because teachers don't have enough time to cover all the important information as it is. I would argue that on the contrary, cramming the maximum possible amount of information into a semester is in fact quite inefficient. Educational psychologists have convincingly demonstrated that material taught in this fashion is not effectively retained by students. More to the point for our purposes, this traditional approach to teaching science drives many potential students away from our field of study by creating a deceptive stripped-down picture of the scientific process. In particular, it appears that women and minorities are disproportionately put off by decontextualized science teaching. Restoring the social, historical, and ethical context of science in the classroom will serve a dual function. By giving a more accurate picture of how the scientific process really works, it will both teach effective scientific thinking and make the pursuit of science more interesting and attractive to traditionally underrepresented groups.

## NOTES

1. Bernice Resnick Sandler, Lisa A. Silverberg, and Robert M. Hall, *The Chilly Classroom Climate: A Guide to Improve the Education of Women* (National Association for Women in Education), p. 34.

2. Quoted in Sandler et al., *The Chilly Classroom Climate*, p. 35.

3. Teresa Arámbula-Greenfield, "Teaching Science Within a Feminist Pedagogical Framework," *Feminist Teacher* vol. 9 no. 3 (Fall/Winter 1995), p. 111.

4. Barbara Brownstein, "Site Visit Report," June 6, 1994, p. 5.

5. See for example Sue V. Rosser, *Female-Friendly Science: Applying Women's Studies Methods and Theories to Attract Students* (New York: Teachers College Press, 1990); Myra Sadker and David Sadker, *Failing at Fairness: How Our Schools Cheat Girls* (New York: Simon and Schuster, 1994).

6. Sheila Tobias, *They're Not Dumb, They're Different: Stalking the Second Tier* (Tucson: Research Corporation, 1990).

7. Marshall Sundberg, Michael Dini, and Elizabeth Li, "Decreasing course content improves student comprehension of science and attitudes towards science in freshman biology," *Journal of Research in Science Teaching* 31 (1994): 679-93.

8. Horace Freeland Judson, *The Eighth Day of Creation: The Makers of the Revolution in Biology* (New York: Simon and Schuster, 1979).

9. See Ruth Bleier, *Science and Gender: A Critique of Biology and Its Theories on Women* (New York: Pergamon Press, 1984); Anne Fausto-Sterling, *Myths of Gender: Biological Theories about Women and Men* (New York: Basic Books, 1992); Stephen Jay Gould, *The Mismeasure of Man* (New York: W. W. Norton and Co., 1981); and Paula Caplan and Jeremy Caplan, *Thinking Critically about Research on Sex and Gender* (New York: Harper & Row, 1994).

10. Sarah Blaffer Hrdy, *The Woman That Never Evolved* (Cambridge: Harvard University Press, 1981); V. Morell, "Seeing Nature through the Lens of Gender," *Science* 260 (1993), pp. 428-29.

*Chapter 5*

# Building Collaborative Communities

**Laura Stempel with Cheryl Ney and Jacqueline Ross**

> Professors, to be fully effective, cannot work continuously in isolation. . . . In the end, scholarship at its best should bring faculty together. A campuswide, collaborative effort around teaching would be mutually enriching. A similar case can be made for cooperative research, as investigators talk increasingly about "networks of knowledge," even as individual creativity is recognized and affirmed. Integrative work, by its very definition, cuts across the disciplines. And in the application of knowledge, the complex social and economic and political problems of our time increasingly require a team approach.
>
> —Ernest L. Boyer, *Scholarship Redefined*[1]

As Jacqueline Ross described in Chapter 1, the development of a collaborative community of scholars, teachers, and students has been key to the success of the Women and Science project. Participants have noted throughout this book that the concept of collaboration not only echoes the way scientists, mathematicians, and engineers actually work, but provides a congenial setting and an effective learning vehicle for students of all kinds—and especially for members of the underrepresented groups targeted by projects like this one. Earlier chapters have emphasized the importance of encouraging specific forms of collaboration among students, including the essential role that can be played in the science classroom by group work, discussion, and discovery-based learning and teaching methods. In this chapter, the emphasis is on the collaborations that were fostered among faculty members at different campuses, and the benefits that this style of working offers to teachers and scholars pursuing curriculum reform, pedagogical innovation, and a more inclusive approach to teaching and learning in the sciences.

As Cheryl Ney pointed out in Chapter 2, a constructivist approach such as the one taken by the UW Women and Science project works best when faculty members are given the opportunity to experience the same kind of discovery method of learning that they will be using in their own classrooms. In the faculty reflections that follow, it is clear that for many project participants, a key moment came when they realized that the faculty development activities in which they were taking part mirrored the classroom activities they were being trained to employ.

Several different levels of collaborative effort are necessary in order for faculty communities like the one described in this chapter to develop. Faculty need to be able to work together within their individual departments as well as across departments, disciplines, and even colleges—between, for instance, liberal arts, where basic math and science courses are offered, and the schools of agriculture, nursing, and engineering whose students take many of those courses. Faculty must also be able to collaborate with their campus administrators and, in cases such as the University of Wisconsin, system administrators. The coordination of effort among project participants, deans, vice presidents, budget managers, and project administrators can be crucial to a project's success, and campus administrators must be able to work with project administrators, project administrators with those from the granting agency, and so forth.

Just as we saw in Chapter 4 that the classroom climate affects how and how well students learn, we will see in this chapter that climate operates at the departmental, campus, and system levels, too. The relative warmth or chilliness of a specific department can have a profound effect on faculty members, and one of the Women and Science project's goals was to improve the climate in which women faculty work by offering them opportunities to build communities across subfields, disciplines, and even campuses. In elaborating on the idea of a collaborative community, we want to underline the notion of faculty as learners as well as teachers. While, in their role as scholars, faculty members often do understand themselves as "students" of their own research specialties, the goal of the Women and Science Program was to foster communities of faculty with teaching as a focus of research. When this kind of shift in research attention occurs, teachers are able to develop a reflective teaching practice, as Cheryl Ney discusses in Chapter 2—and such a transformation needs to happen beyond the confines of schools of education where theories of teaching and learning are conventionally debated.

The idea that faculty members are scholars about teaching is, in fact, central to the project's practice of bringing Distinguished Visiting Professors (DVPs) to campuses around the state. The project created short-term positions for visitors whose entire responsibility was to get people to focus on, rethink, and thus revitalize their teaching, a revitalization that can be seen in the pedagogical innovations and experiments described in Chapter 3. This chapter shows how such faculty collaborations can be organized and implemented, and what its costs and benefits may be for faculty, departments, and campuses. In describing these features of the project, we also consider the barriers to collaboration faced throughout, and how participants overcame them.

Behind the project's collaborations among faculty, administrators, and staff on individual campuses and across the statewide university system lie the same principles as those supporting the idea of students working together. This is how science is actually practiced: Scientists work together in research, in contrast to the antiquated cultural stereotype of isolated "geniuses" in their individual labs; and people, including researchers and teachers themselves, learn better when they learn by doing.

As we have repeatedly noted, the Women and Science Program assumes that what works for women students works for male students as well, and in the same way, what works for students works for their teachers. Not only do faculty members benefit from experiencing in their own development activities the kind of learning process their students will have, they also become invested in the project by collaborating to ensure its continuing success. In a project like this one, members of the program community have to sustain its aims by, among other things, being trained to be and then working as consultants to other constituencies, so the project must first be "owned" by the participants, rather than simply being imposed on them from above. This process of coming to ownership is a gradual one and requires that administrators listen to participants, take all of their concerns seriously, and work together with them to incorporate any suggestions that seemed likely to improve the program. In this particular project, that process was helped by the fact that evaluators made periodic reports to the project administrators with suggestions for improvement, rather than waiting until the project's end to issue a concluding analysis. In order to accomplish real collaboration, projects and tasks in the collaborative setting must be crafted to emphasize the interdependence of the participants on one another. Project completion depends on the work of the entire group and not any individual member of the group. Furthermore, groups don't reach the stage of this kind of interdependent activity until they have spent a considerable amount of time working together; it takes time for participants to get to know each other well enough to cooperate effectively. Organizers of collaborative faculty development activities must thus construct them with close attention to these features of collaboration.

Yet making collaboration work is a difficult endeavor, and even a fundamental commitment to the goals of a project like this one does not guarantee that the work will be easy. The Women and Science Program has faced many challenges, some of them quite specific to the structure of the UW System and others of a more general nature, issues that are likely to arise in any project that asks faculty, staff, and administrators to rethink some of their most basic premises about how science learning, teaching, and faculty development should be conducted. Throughout the project, there have been problems involving communication and organization, some attributable to the novelty of the undertaking, others to simple misunderstanding, and still others due to active resistance on the part of people opposed to the program's goals. This chapter describes several different levels of collaboration that occurred during the Women and Science project, and at each level, we outline both the success and difficulties. Although these have occurred in a large state system, they have involved a variety of institutions of different sizes, from small two-year campuses to comprehensive universities to large research institutions. Similar obstacles might occur in projects carried out at both public and private institutions, on a single campus, or even within a single department.

In both the development and beginning phases of these collaborations, it is important to identify a critical mass of people interested in the project's goals and to build on the specific disciplinary, pedagogical, and other interests that bring them to the collaboration. That critical mass may come from a faculty/administration coalition, as well as from coalitions among faculty across departments (such as science and math) or even across the divide between science and the humanities. As these examples suggest, it is important to build coalitions at as many levels as possible, since the more support a project has, the greater the chance for its long-term success. Yet it is also crucial to remember that coalition-building and collaboration can take place even without support at every level, so that, for instance, if a dean or department chair is not supportive, the project can still be developed based on the commitment and initiative of a critical mass of faculty. In addition, while the project goals focus on student learning, everyone involved also needs to understand that participation does not necessarily require that all members agree on all the details of implementing those goals.

The collaborative communities on which projects like this one depend need to be both fostered and continually cultivated, not simply through discussion among members, but through feedback, rewards, and opportunities for reflection. The Women and Science project benefited from the qualitative assessments performed by project evaluators who interviewed participants, thus providing faculty, administrators, and others involved in the program a chance to think in a focused and reflective way on what they had been doing. Equally important, the evaluators presented their assessments through interim reports to project administrators, giving participants yet another opportunity to think concretely about their progress toward project goals.

In the case of this project, specific resources were available to help foster both collaboration and a sense of community among participants. For example, the project and individual campuses were able to support two-day workshops and pay for participants' accommodations. These lengthy workshops allowed project members the opportunity to get to know one another socially as well as to learn from fellow participants in an academic setting. However, not all rewards require this kind of financial support. Public acknowledgment and praise from campus and project administrators, and letters for participants' personnel files, can help to assure that faculty and staff get credit for the contributions. Administrators' direct involvement can also keep participants on task and sticking to deadlines, both of which are difficult in the midst of teaching. For DVPs—who were, after all, outsiders at the campuses where they worked—involvement with administrators provided information that helped them to understand politics and other contextual issues on campus and within the project as a whole.

The Women and Science project itself was a collective one from its beginnings, and involved the creation of a large, statewide collaborative community that included faculty, campus and System administrators, staff members, and the individuals directly appointed to carry out and administer the program. That community met through conferences and retreats, which will be described at the end of this chapter, and program staffers communicated regularly with other participants via mailings, phone calls, and campus visits. The collaboration also included people not formally associated with the program, such as UW System Women's Studies Librarian Phyllis Holman Weisbard, who attended retreats and conferences, provided resources to participants, and even consulted on the literature review for this book. But most of the work of curriculum reform and faculty development occurred in the smaller communities created on individual campuses and multi-campus sites, facilitated by the presence of the DVPs whose work with faculty colleagues placed collaboration at the core of the entire undertaking.

## THE COLLABORATIVE COMMUNITY

Faculty and staff throughout the state, including those at single-campus sites, undertook a common set of collaborative experiences in a project that often required them to work together across departmental and disciplinary boundaries. But one of the project's sites—the three-campus Collaborative Community consisting of UW-River Falls, UW-Eau Claire, and UW-Stout (which are from 30 to 77 miles apart)—provides a particularly striking demonstration of both the difficulties and the rewards of this kind of collaboration. This multi-campus site required DVPs, Faculty Fellows, and other participants to cooperate across geographical distance and administrative barriers as well, and challenged the DVPs and their Campus Coordinators to find ways to overcome differences and divisions between campuses. Yet, while the specific logistical obstacles, such as figuring out how to gather participants who work on separate campuses, may be unusual, the fact that DVPs and administrators were able to overcome these successfully enough to effect real change on the campuses where they worked testifies to both the power and the importance of building collaborative communities among faculty in the sciences.

The three-campus collaboration meant that three separate Campus Coordinators worked together on a project that eventually involved, according to Sheue L. Keenan (UW-River Falls), "three universities, fourteen departments, and about sixty science and mathematics faculty." In fact, because of the complexities of working with three campuses, the role of Campus Coordinator developed far beyond the original project proposal. Their successful collaboration with each other and with the program participants was so essential to the success of the Collaborative Community itself that this model was used in all subsequent phases of the project, even at single-campus sites. Just as the DVPs were focused on helping faculty members to revitalize their teaching, the coordinators' roles were focused on organizing and carrying out the project at the local level, and their roles evolved as the project did. Among other contributions, the coordinators served as both institutional and personal hosts to the DVPs, and thus relieved a lot of the burden of management from those visiting scholar/teachers, who were therefore free to concentrate on their teaching and modeling roles as well as on the task of building faculty communities.

In Appendix A, the project evaluators use the Collaborative Community as a case study, analyzing faculty and student responses to the activities comprising the Women and Science Program. Here, however, we use it as a focus in another way, arguing that the practical and administrative complications facing this multi-campus site simply clarify the challenges confronting everyone who undertakes a project like this one. The solutions devised by Collaborative Community members were later replicated at other campuses and, indeed, throughout the statewide project, and their experiences can provide a model for collaborators in projects at other campuses as well.

## The DVP Experience: Sherrie Nicol

For Sherrie Nicol—the Collaborative Community's first Distinguished Visiting Professor and the first one at any campus to be in residence at a project site—the goal of systemic change required access to a larger audience than the individual participants committed to the program:

> The goal of increasing participation by women and minority students in the sciences demands a program which reaches many faculty. Working with the Faculty Fellows and having [the usual] three Saturday workshops would not be sufficient in promoting systematic change in the undergraduate curriculum, the climate in the science disciplines, and teaching strategies. While the enthusiasm of the Fellows could spread within their own departments, that would still leave many disciplines on each campus out of the mainstream of the program. I felt that larger cross-discipline communities on each campus, focusing on sharing and developing their teaching strategies, would be more productive. .. .
>
> Given the short distance between these campuses, I anticipated discipline-specific communities already to be in existence. The truth, however, was that . . . these campuses might have been 500 miles apart. . . . Saturday workshops would provide a medium for community growth within disciplines and across campuses.

Nicol traveled regularly among the three campuses, conducting workshops every few weeks that, she writes, "provided faculty from a variety of disciplines the opportunity to network with others also interested in teaching changes." There were obstacles: Some faculty members resisted any direct references to the study of women, or to research from Women's Studies, objections Nicol met by using titles for the seminars that were neutral and did not explicitly mention women. There was a fundamental disbelief among some faculty—even those committed to the project's goals—that Nicol, a mathematician, could have anything useful to teach to, say, a chemist. There were also organizational problems, such as serious confusion about the definition and scheduling of release-time, and even about the goals of the program itself—problems that required that Nicol meet with administrators at all three campuses to answer questions and explain the project more fully. Yet, she writes, "at each campus small communities were forming and the seminars provided a regular meeting for these groups."

University of Wisconsin-River Falls Dean Neal Prochnow, a member of the project's policy Advisory Board and himself a physics educator, notes that "it was important that the DVPs were individuals external to the three institutions involved in the collaborative [community] . . . [because they were thus] able to provide the 'outside experts'' perspective of the curriculum as well as the faculty development activities." According to campus coordinator Sheue Keenan, however, the project faced difficulties that could not be solved simply by an appeal to a DVP's expertise, and she credits Prochnow's support for lending additional credibility to the undertaking: "A few faculty . . . shied away from the program because they thought 'Women and Science' implied that men were not welcome. This stigma continued to plague the program, although it lessened as time went on." (Indeed, one faculty member interviewed by project evaluators Gloria Rogers and Judith Levy indicated discomfort with the premises of the program: He "was originally involved but then dropped out of the project [and] indicated that he did not know what the goals of the program were when he originally agreed to participate."[2])

Many of the challenges the Collaborative Community faced were an innocent function of the complexity of a statewide project like this one, but some, such as communication problems, often seemed to mask resistance to the program's goals. On campuses where there was support for the program at both administrative and faculty levels, there were no serious complaints about "poor communication," whereas on other campuses, no matter how many times particular pieces of information were repeated in person and in writing, the same people claimed never to have heard them before. One lesson for others undertaking similar projects, then, is to anticipate that complaints about communication may be a site for expressing other objections, and to pre-empt serious problems by producing extensive, multiple forms of communication to faculty, administrators, advisory board members, and other participants, and by making all provision of information redundant through email, letters with copies, campus visits, and so on.

## THE DVP EXPERIENCE: CHERYL NEY

By the time Cheryl Ney, the Community's second Distinguished Visiting Professor, arrived, many of the initial organizational difficulties had been reduced, yet, write evaluators Rogers and Levy:

> A meeting with department chairs and deans at one campus had overtones of defensiveness and resistance. Concern was expressed [by those interviewed for the interim report]. . . that the program got off to a bad start and the project was still under a lot of criticism. One of the department chairs who had been characterized as resistant was critical that the presence of the program implied that something was wrong and that there had been no assessment to see if that was, in fact, true.

In fact, the project's goal of revitalizing teaching does not necessarily imply that "something is wrong," but assumes that a fresh approach, an outsider's expertise, and an opportunity to concentrate on pedagogical concerns can benefit every faculty member. In addition, regardless of claims that "there had been no assessment" of a specific problem in need of a solution, the project's attention to the attraction and retention of so-called "second tier" students was intended to address a problem well supported by research.[3]

Ney did, however, face a contentious curriculum issue shortly after her arrival for her residency at UW-River Falls (from which she also traveled extensively among the three campuses). She was scheduled to teach General Chemistry and its accompanying lab, but a number of department members were resistant to her plan of using a lab manual of her choice, rather than the one produced in-house. However, as Campus Coordinator Sheue Keenan reports, the controversy over Ney's introduction of a different lab manual merely brought to the surface a conflict that already existed within the department:

> There were discussions in the department for a long time about the need to revise our general chemistry manual to include more discovery-based experiments. The chemistry department had been using a very traditional (verification-based) . . . manual, edited by a senior faculty member with little input from the others. Some felt strongly that it was time to examine our pedagogical approach in teaching our laboratories. Cheryl's presence helped to bring the issue to the top of the departmental agenda. After many hours of heated debate, the department adopted a procedure by which a faculty member can contribute to the writing and editing of the manual. It was also agreed that flexibility be allowed in teaching the laboratories, so that each faculty [member] could test one or two new experiments or try new teaching strategies.

In her own report on her experience as a DVP, Ney describes other incidents that indicated a degree of local hostility to the program and to women in the sciences more generally: "In one department, a faculty member told me that chemistry was no place for a woman since it might harm her capacity to reproduce or endanger a pregnancy. In another, it was politically risky for a woman faculty member to be seen with me. Action was taken against a program participant in a retention vote in a third department." Ney also witnessed what she calls "covert chilly behavior," such as departmental decisions being made in hallway discussions that excluded female faculty members.

These expressions of hostility are obviously crucial in assessing the larger climate affecting women in the sciences, but in the specific context of collaborative projects like this one, they matter in another way as well: "My sense . . . is that it has been difficult for women in science to acknowledge the climate they find themselves in," writes Ney, yet that climate makes many women reluctant to work in the very projects that might help alleviate the problems they face. "How can we promote collaborative learning in the classroom when we have difficulty collaborating ourselves?" Ney's experience illustrates once again the value of having project DVPs from outside of the campuses at which they worked. As external consultants, they are able not only to measure the climate women faculty face, but can also bring together women who feel isolated or are, because of that climate, reluctant to voice their concerns.

Like Nicol before her, Ney worked to promote the notion of faculty as scholars of teaching through a variety of development activities, which she organized according to the constructivist model she has described in Chapter 2. The fact that her "students" were actually teachers presented special challenges to building the faculty community:

> Faculty . . . had a wide range of background understandings about teaching and learning from their own practice, as well as the scholarship in education and women's studies. This often made communication difficult. In addition there was the difficulty of speaking across disciplines, including the subdisciplines of science.
>
> I could sum up the experience of working with these learners as being more difficult than any classroom situation I had previously encountered. Their motivations, epistemological commitments, and background knowledge were more varied than in a classroom situation. In addition, I found it very difficult to build the kind of community that is possible to establish in the classroom.

## FACULTY FELLOWS AS STUDENTS

For Distinguished Visiting Professor Sherrie Nicol, modeling different teaching methodologies for Faculty Fellows at three different campuses working in four different disciplines represented a major logistical challenge. "Four themes emerged," she writes of her initial reflections on her own teaching style:

> anxiety-reducing strategies, group work, classroom material relevant to students, and ownership (discovery/responsibility) development. . . . Although my initial framework was for all the Faculty Fellows to observe my class once a week with a discussion period to follow, this was not feasible due to time and travel constraints. . . . [Instead,] I decided to have the Faculty Fellows work in small groups to develop mini-workshops focusing on the four themes. . . . In doing this they would certainly experience anxiety in preparing a mini-workshop on a teaching strategy with which they were not familiar. The entire exercise would be relevant to their profession as teachers. Finally, they would develop ownership over the theme by discovering how to utilize [it] in their own discipline and present it to faculty in other disciplines. It was not until the Fellows repeated their refined mini-workshops [two months later] . . . that a few of [them] realized that they were learning about teaching in exactly the same ways the four themes pertained to students learning in a discipline.

As a DVP in the same collaborative community, Cheryl Ney also introduced her colleagues to a variety of different classroom practices, and her experience provides a useful summary of the potential benefits of the kinds of pedagogical change described in Chapter 3, as well as a realistic acknowledgment that such change does not always come easily. In her assessment of her tenure in the project, Ney notes that, despite the obstacles they faced, faculty members seemed gradually to grow more comfortable with these innovations, yet she also acknowledges that these student-centered methods demand more faculty time and effort than other approaches:

> It is my sense that in general, faculty [at the campuses she visited] are feeling more confident about teaching intuitively and developing a reflective teaching practice. . . . They seem to be developing activities, assignments, course delivery and course structure that makes sense to them from their past teaching experiences and from their present ones. Many faculty seem to be listening more actively to the students' experience of learning through journal and classroom assessments. . . . I see many faculty examining students' work as data for their own research questions on teaching and learning. . . . As we move to student-centered approaches and begin to recognize the differences among our students, we work toward tailoring our courses in ways that were not required in the past. The payoff with this approach is large—students seem empowered, teaching is more exciting, but the costs must be recognized. Some faculty are already thinking about how scheduling and courses could be restructured to accomodate student-centered learning.

* * * * *

As Sherrie Nicol discovered, participants from different disciplines often needed to be persuaded that someone with different training and expertise could help them improve their own teaching. Although the Women and Science project is united by a broad multidisciplinary rubric, ideas about how to improve teaching have not always crossed fields easily, and teachers in, say, math have faced different problems than those in computer science or chemistry. Commenting on Cheryl Ney's visit to the UW-Eau Claire campus, mathematician Alex Smith describes Ney's epistemological dilemma from the learner's side:

> It often seemed that the mathematicians could not understand this chemist and that this chemist could not understand the mathematicians. One important reason for this initial failure to connect is that mathematical reality and physical reality are probably quite different. Mathematicians might understand that when a chemist speaks of a "model" for the foundations of a science, they are speaking of something like the Ptolemaic Model of the Universe, which eventually was replaced by the Copernican Model which was eventually replaced by the Big Bang Model, etc. Models for the foundations of mathematics include philosophies of mathematics such as formalism, mathematical platonism, and intuitionism—which is often called constructivism [the theoretical approach Ney advocates for pedagogical reform] . . . the suggestion that constructivism might provide theoretical foundations for our reform efforts was utterly confusing because we had incorrectly equated it with intuitionism/constructivism.

Yet the project's cross-disciplinary discussions have also inspired many participants. As we saw in Chapter 3, faculty members familiar with feminist pedagogies often adapted them for the science classroom, and such collaboration between colleagues who are accustomed to seeing themselves as separated by specialty is perhaps one of academic feminism's most important legacies to projects like this one.

Ney had many positive experiences with her community of faculty learners, and found that frequent informal meetings (such as pizza parties and hallway conversations) and a clear sense of shared goals seemed to make an enormous difference: "The most effective activities . . . seemed to be those where I was working directly with faculty who were teaching a course that I was teaching. Sharing an office with a faculty person teaching another section of the same course was . . . the most effective opportunity for change—for both of us. The opportunity for collaborative work on courses is rare." She also found that tandem teaching with a Faculty Fellow was extremely effective.

She reports that the final workshop of the semester, which was a showcase for the NSF evaluators, was especially satisfying, perhaps because participants had finally gotten to know each other well enough to collaborate effectively. For Ney, the success of that workshop was also a good exhibition of collaboration at every level of the project, from students (who were interviewed by the evaluators)[4] through faculty, advisory board members, and administrators, and she was especially pleased by the idea that participants—and students in particular—could speak directly to federal representatives, in the person of the NSF evaluators, about the successes of the program.

It is clear from the evaluators' interim report that the Collaborative Community had important positive effects. Some female faculty members who felt that departmental colleagues didn't support them found a sense of community by visiting one of the other two campuses. "One faculty member reported that friendships had grown . . . . 'Cheryl (Ney) put us together with informal pizza parties rotated from place to place. It created a safe place for women to talk.'" But, the evaluators add, participants realize that the kind of collaboration exemplified by this project requires "total commitment in order to be successful. It has worked in the Collaborative Community because [they] have been willing to work hard, compromise, and have a sense of humor" ("Interim Report," p. 15).

## Faculty Fellows' Perspective

The Faculty Fellows were key members of the Women and Science project, since they were the ones who would stay on campus after its completion, teach the new and revamped courses, and implement other curricular changes within their own departments and through consulting visits to others. Fellows were recruited from among campus faculty, and although project planners had originally wanted tenured participants, since participation in a program like this—with its emphasis on teaching and not on disciplinary expertise—could jeopardize promotion decisions for untenured faculty. Furthermore, tenured faculty represent permanence for a curriculum, while tenure status usually allows a faculty member to have more influence over others in the department. However, mostly untenured faculty and staff applied for the positions, although as the project progressed, more and more tenured faculty became involved.

One participant in the three-campus site was Faculty Fellow and mathematician Loretta Robb Thielman (UW-Stout), who notes that she had long wanted to introduce new collaborative teaching strategies into her courses, even attending conferences on pedagogical issues:

> I wanted my students to be active learners who actually appreciated that mathematics or statistics I was trying to teach. Unfortunately, my inspiration and resolution to try new methodologies always faded under the pressures of time and other responsibilities. I felt I really needed TIME to think about incorporating new teaching techniques into my classes. Having a support group or

> even one other person with similar interests to keep me going would have been even better. Best of all would have been having an expert teacher providing a working model right in my department, available for my observation as well as serving as a coach who could get me started and keep my on track. This program proposed just what I wanted for myself—the opportunity to become a more effective teacher.

When Distinguished Visiting Professor Sherrie Nicol assigned the Faculty Fellows to run mini-workshops in October, Thielman chose one on Cooperative Learning, which she organized with Fellow Brian Bansenauer (UW-Eau Claire), also a math educator, and it was this part of the project that showed her the real benefits of collaboration:

> It was . . . in the middle of the . . . mini-workshop when the [faculty] cooperative groups were working away, when it suddenly dawned on me that Sherrie had used the most effective teaching techniques on us Faculty Fellows from day one: she assigned us the task [of] choosing a topic and forming cooperative groups, providing us with reference materials and encouragement; . . . we assumed ownership for our topic; . . . and we experienced the power of cooperative problem-solving. I suddenly saw the cooperative learning experience from the student's point of view.

Faculty Fellow Rhonda Scott-Ennis also reported success for the three-campus collaboration, but she cautions that the work they accomplished during the Women and Science project is not enough to transform the science curriculum. Referring to a model of curricular change that posits three distinct phases (exploration, the introduction of new terms, and application), she writes:

> While our workshops have been positively received, they are only a small part of the process needed in bringing about institutional change in introductory curricula. Workshops are an effective means of making faculty aware of new theories of teaching and facilitating the initial exploration phase. However, curricular change requires that all of the phases be carried out at any campus wishing to incorporate these new teaching strategies that have been shown to benefit all students, but particular women and minorities. One of the strengths of our experience was the constant presence and opportunity to work closely with a DVP for three semesters, as well as the collaboration between Barb, Kim, and myself as we worked through this process together. . . . [But] faculty need more opportunities to meet with other faculty who are interested in curricular change to brainstorm new ideas and to encourage each other as new ideas are tried.

## A Campus Administrator's Perspective

Neal Prochnow, the UW-River Falls dean whose support for the development of the Collaborative Community was seen by many as crucial to its success, notes that, "as with any project, there was a need to develop tangible, concrete projects." In this case, the tangible products at the UW-River Falls campus included greater direct involvement of science and math faculty in Women's Studies, a new course on "Gender Issues in Science" (see Appendix), a revised alternative laboratory manual for the introductory Chemistry course, and ongoing "mini-workshop modules that illustrate specific teaching strategies and can be used for faculty development activities." Yet the most important outcome, according to Prochnow, was not one of these tangible products, but a less easily documented achievement:

> . . . acquiring knowledge is a collaborative social process. . . . [S]tudents who study together in groups do better than those who do not. Faculty who work together will do better at accepting new ideas with respect to teaching and integrating these ideas into their teaching philosophy and style than faculty working alone or within a single department. . . . [T]he presence of Dr. Ney on the campus increased the social aspects of teaching and the residual benefit has been an informal support network among those involved in the project as well as greater acceptance of new ideas with respect to teaching. This intangible product will have a lasting impact on students.

Still, writes Prochnow, the project's success "was not without the tension associated with change." Better early communication and the retention of a project title focused on revitalizing introductory curricula might, he thinks, have decreased that tension, yet he acknowledges that curricular change takes time and requires not only research on learning, but forums that go beyond workshops and seminars. "It is in the informal conversations where the philosophical aspects of acquiring knowledge can be openly discussed. . . . I can honestly state that the numerous informal conversations that I had with Dr. Ney resulted in an adjustment of my philosophical approach to teaching."

*****

# A Project Administrator's Perspective

**Jacqueline Ross**

While I am a faculty member as well as an administrator, I am not a scientist and was not, until my involvement with this project, accustomed to working within the scientific academic culture. I did not quite realize, at the beginning, the magnitude of the task involved in developing and implementing a program aimed at promoting changes in the content, pedagogy, and climate of the sciences. As a result, initially at least, I underestimated the degree of resistance and made a number of mistakes. For example, it did not occur to me that when a department chair told me enthusiastically that his faculty were thrilled by the selection of their program as a host community for a DVP, he was dissembling. Thus, I was very surprised to learn shortly before the beginning of the semester, when the visit was to occur, that the department knew nothing about it. And when they did find out, they were hostile and not inclined to be receptive to their guest or to the program administrators. Yet from experiences like thus, difficult though they may have been at the time, I learned a great deal about the importance of effective and repeated communication.

A major lesson was the importance of developing partnerships between the project administrators, the faculty, and staff in the host communities, and, as the project matured, with other Women and Science communities. Second, it was crucial that the participants take ownership of the project. This is a very gradual process, but it must occur. While there were commonalities among the host communities, each also had its distinct culture and each had to be able to adapt the program model to meet the needs and interests of its own campus. This may seem self-evident, but it can be a hard lesson for project administrators who must listen and be responsive to participants, incorporating suggestions for improving the program.[5] If the administrators are flexible in applying and adapting the program model, they are likely to forge stronger partnerships with the campus-based participants. At the same time, our goal was for the Systemwide community, including the Advisory Board, to take ownership and serve as the ambassadors to the campuses. As Chapter 6 indicates, this process eventually led to the institutionalization of the program in the UW System.

Based on our experience, including mistakes that we made, I have some suggestions for anyone embarking on a project such as this, whatever the scale. Once we realized the importance of communication in forging partnerships and encouraging ownership, we arranged for conference calls with the DVP and key members of the host community as long as a year in advance. As the time of the visit grew nearer and at the beginning of the semester, we scheduled set-up meetings that included presentations by the DVP to faculty, staff, and department chairs, as well as deans, provosts, and chancellors.

Usually, these groups gathered in different meetings, depending on what seemed most appropriate for each campus. Throughout the semester, we maintained contact with these groups as well as with the DVP and Fellows. At the conclusion of the DVP's visit, we met once again with these groups to solicit their response to and suggestions for the project. Finally, each DVP prepared a written final report on her visit, which she presented during debriefing sessions with campus and UW System administrators. While some of these activities may seem redundant, the number and frequency of the communications involved were, I believe, instrumental in the reinforcing the place of the program in the campus and system.

This discussion of the three-campus Collaborative Community provides a case study of the issues involved in the Women and Science project as a whole. Coordinating this kind of collaborative program across a large statewide university system presents special challenges because of geographical distance, administrative structure, and differences in student populations and campus missions among individual campuses within the system. But many of the solutions devised to bridge the gaps among collaborators on different campuses can be applied to similar collaborations in other kinds of institutions, and even on single campuses or within individual departments. These solutions also suggest some of the strategies that can be used to foster collaboration and communication that go beyond specific projects, particularly among institutions involved in separate programs of curricular reform and faculty development. The idea of a collaborative community can even be expanded to include other groups of educators and administrators committed to the kinds of changes designed to improve the campus climate for women in science, mathematics, and engineering, and to attract to and retain in those fields "second-tier" students and those from underrepresented groups. The lessons of the UW program are as applicable to that nationwide community as they are to faculty and administrators undertaking local reform projects.

*****

## Conference and Retreats

One challenge in a large project like this one is to make participants feel that they are all part of the same undertaking and are working toward the same ends. Project-wide conferences and retreats, while not part of the original project proposal, proved to be extremely valuable, giving participants several opportunities to spend time together for concentrated work, and providing the kinds of informal contacts that proved to be crucial to advancing the program's goals. The kick-off conference, for instance, introduced UW System administrators, faculty, and staff to the goals and objectives of the Women and Science project. Sheila Tobias, whose work on the phenomenon of "second tier" students was a fundamental component of the project's design, served as the keynote speaker, with faculty from within and outside of the UW System describing specific classroom innovations and introducing programs aimed at retaining women and minorities in the sciences. Some conference attendees expressed uncertainty about the availability of funding for curriculum reform and faculty development activities like the ones presenters described, and more than half were concerned that their home departments appeared uninterested in the problems raised by the conference, commenting that speakers like Tobias underestimated the degree of resistance they faced in trying to implement major reforms. Most, however, were enthusiastic about the conference itself, especially the small-group and workshop discussions, and some went home inspired to begin a concerted effort on their own campuses.

In April 1996, the NSF-funded period ended with another conference, "An End and a Beginning: Science, Diversity and Community," which also drew participants from around the state. Writes Nancy Mortell, assistant to the program director:

> The conference was designed to demonstrate successful and innovative models of pedagogy and curriculum reform and faculty development that the Women and Science Program had developed which: a) improve the overall quality and effectiveness of introductory science teaching; b) increase the number of women and minority students attracted to and retained in science educational programs; and c) transform overall curricula and climate.
>
> At the conference, former DVPs, Faculty Fellows, and other faculty and administrators involved in the project conducted seminars and workshops . . . on gender, race, and science issues pertaining to pedagogy, climate, and curricular reform, enabling presenters to demonstrate innovative teaching strategies and techniques they had developed in their classrooms as well as to share with other conference participants the successes and challenges they faced in introducing new approaches in the classroom. The conference also included a panel discussion on "Institutional Leadership and Change" as well as [preliminary results] presented by the project's evaluation team. Through these workshops, presentations, and panel discussions, the conference aimed to encourage and facilitate UW System administrators' use of the Women and Science model for faculty development to achieve curriculum reform and to recruit and retain a more diverse group of qualified students in the sciences.
>
> The conference also included planning sessions to assist campuses in identifying strategies they could use to implement the goals of the program at their home institutions. Program participants and former DVPs facilitated these sessions, in which representatives from each institution began to outline the steps necessary to incorporate changes into their curriculum, pedagogy, and/or faculty development efforts; and identified potential problems and suggested methods of overcoming barriers to incorporating these changes. Campus planning session participants also agreed to schedule a follow-up meeting at their home institution to further address the above issues and objectives.

Participants responded enthusiastically in their evaluations of this closing conference, especially in describing the project's larger implications. One wrote, for example, about the "unique opportunity to extend the impact of this NSF grant by working through institutional continuing education office[s] to extend the knowledge and energy to girls and women in our communities, K-12 educators and technical/science related business. Let's not miss it!" Another noted the persuasiveness of this project's results: "Even as a woman in science (chemistry) myself, I enter workshops/conferences on women in science with some criticism and skepticism. The single most effective thing I heard all day was [evaluator] Gloria Rogers' presentation on the assessment of this program. Her data on the change in student (male and female) perceptions of their own competence and attitudes has convinced me that this effort is worthwhile."

Retreats were also part of the Women and Science Program's gradual evolution. Once again, Nancy Mortell describes their development and purposes:

> In the spring of 1994, the Women and Science Program administrators decided to sponsor a retreat which enabled project participants from dispersed institutions to get together face-to-face and discuss their experiences with the project. The initial retreat was so successful that the project's administrators decided to make it an annual event.
>
> Women and Science Program retreats . . . facilitated the development of a state-wide women and science community. Engaging in informal discussions with other participants provided [everyone] with an opportunity to share their experiences, strategies, methods, challenges, and success. Through informal discussions, participants developed networks which sustained and supported them throughout the year. As the program expanded to new institutions and involved new DVPs, administrators, faculty, and staff, the retreats served to acquaint these new [members] with the program and helped expand the women and science community.
>
> Retreats also provided participants individually or as teams with an opportunity to practice presentations and workshops that they had developed [in a friendly and supportive environment], and to formally discuss new strategies for the classroom. . . . Presenters then received feedback from program participants and former DVPs which enabled them to modify and improve their presentation techniques, strategies, and/or approaches.
>
> Additionally, retreats featured guest speakers, workshops, and seminars on successful innovations participants had introduced in their classrooms and institutions. At one retreat, for example, the director of a UW Women's Studies program led a discussion on forming links between women's studies/feminism and the sciences, teaching, and learning, while at another, a former DVP conducted a seminar on epistemological issues and science.

> Retreats also provided program participants with an opportunity to reflect on the goals, challenges, and successes of the project, as well as the [chance] to plan future directions for the project and develop strategies for institutionalization. Participants were able to meet with the project's evaluation team to discuss the progress of the program to date and to explore possible problem areas and modifications. Retreat participants discussed and identified ways to disseminate and market the project around the state, such as the development of a program brochure and Web page.
>
> Participants put together teams to serve as consultants to other groups and institutions. Additionally, they discussed institutional aspects and issues, such as overcoming administrative barriers and budget constraints as well as the role of tenured faculty in the institutionalization of the program.

## Conclusion

Collaborative communities that are centered around teaching and its scholarship have a significant impact on the ongoing development of faculty, trained as disciplinary experts, in their role as teachers. With intentional and extensive efforts, such communities can and should be fostered. These efforts often require a rethinking of traditional roles, with some faculty in the role of learners (Faculty Fellows), others as managers (Campus Coordinators), and still others as leaders (DVPs, Faculty Fellows developed as trainers). Significantly, administrators are also cast in a variety of roles as managers, creative consultants, and when participating in DVP workshops, as learners about teaching as well. What makes this degree of activity possible seems to be a shared commitment on the part of all involved to the betterment of teaching and the role it plays in enhancing learning for more students.

There is a significant body of scholarship on collaborative learning in the classroom that can be used in the cultivation of interdisciplinary faculty collaborations as well. Faculty who are themselves working in collaboration have an enhanced understanding of the process, and this impacts the nature of the collaborative activities they craft for the classroom. More importantly, when students understand that faculty—people with whom they are acquainted—work together on various projects, they come to understand that knowledge is something developed by people who bring to that process their own different perspectives and talents. Faculty collaboration not only enriches the experience of scholar/teachers, but also provides students with models of their collaborative undertakings in the classroom.

## Notes

1. Ernest L. Boyer, *Scholarship Redefined: Priorities of the Professoriate*, A Special Report from the Carnegie Foundation for the Advancement of Teaching, 5 Ivy Lane, Princeton N.J., p. 75; available from Princeton University Press.

2. Judith Levy and Gloria Rogers, "Interim Evaluation Report," Spring 1993-Spring 1994, p 10; subsequent quotations cited in text as "Interim Report."

3. Sheila Tobias, *They're Not Dumb, They're Different: Stalking the Second Tier*. Tucson: Research Corporation, 1990.

4. While we lack long-term data on the effectiveness of the program in regard to student learning, the evaluators' analysis of their data suggests that positive changes occurred in the attitudes of the students, and in particular women, in the experimental course sections (see Appendix A). As a result, students were more responsive in class, more open to the subject matter, and more willing to take responsibility for participating in the learning process. The evaluators further reported that, in interviews and focus groups, students spoke enthusiastically about the group work and other approaches used by the DVPs and Faculty Fellows. At least one student complained that he was at a disadvantage in being in one of the sections taught by faculty *not* involved in the program. Barbara Brownstein, then of the National Science Foundation, indicated in her report of her 1994 visit to the Collaborative Community: "I should note that the students (selected by Karolyn [Eisenstein, also from NSF] and me from those currently taking the "experimental" classes of the DVP's and Faculty Fellows, were overwhelmingly supportive of the changes and very excited. Several were taking the introductory chemistry class for the second or third time and reported that they "hated chemistry and all science" until now, and some of the initially-hostile students (equally divided between males and females) were now planning to take more science or math." More recent reports of student responses in classes taught by program participants substantiate these preliminary reactions.

5. We were also fortunate in having the counsel of our evaluators throughout the project. However, every project should have an assessment plan, well designed and effectively carried out, to serve this purpose.

*Chapter 6*

# Institutionalizing Change

**Laura Stempel with Cheryl Ney and Jacqueline Ross**

A major problem faced by any successful project is continuing after the pilot period is over. The university must "buy in" to the program, in more than one sense of the term, to effect institutionalization. The innovative strategies that were developed to meet the challenges posed by the three-campus Collaborative Community proved to be critical for the ongoing work of the Women and Science Program. Project members at campuses such as UW-Stevens Point and UW-LaCrosse, whose DVP visits came later, were able to apply many of them at their sites. Equally important, the successes of the Collaborative Community demonstrated that the design of the Women and Science Program was an effective one and therefore could be the foundation of the project's continuation. In the words of NSF evaluator Barbara Brownstein's report on her visit to the site,

> the Collaborative Community . . . is having a major impact on the campuses. The new approaches to teaching of the introductory science courses has certainly won the approval of the students and a cadre of faculty. Many of the faculty, and the academic administration, commented on the impact and the opportunity, many noting that while this program was designed to enhance the success of women in science, the entire student community—male and female, potential science majors and non-majors—were benefitting. The impact on the women faculty, few though they be, was especially important. In group and private discussions the women of the faculty repeated what we hear throughout academia—they are few, frequently isolated, often the junior and most vulnerable members of their departments. In these schools, gathering around their distinguished visitors and carrying the status that a major NSF grant confers, they have established a community. Whether it will sustain them in the long run is not known, but they certainly have a better chance for success in their disciplines and their departments than they had before this program.[1]

This kind of enthusiastic response from one of the program's NSF evaluators, along with the enthusiasm the Community's successes generated within the project's System-wide Advisory Board, helped to sustain the momentum for later work.

Early on, efforts were begun throughout the UW System to institutionalize the changes introduced by the project in order to make them permanent and widespread, and to build a broader collaborative network, especially with Women's Studies programs. In 1994 (about halfway through the project), the theme of the UW System's annual Women's Studies conference was women and science, an indication that even faculty and administrators who were not working directly in the sciences were nevertheless beginning to think of this focus as an important one. The conference also helped project participants, some of whom had never before been actively involved in Women's Studies, to make additional contacts with faculty and staff committed to feminist teaching and scholarship. (One of them even went on to become Women's Studies coordinator on her own campus.)

This chapter explores both local and Systemwide efforts to institutionalize the changes begun in the Women and Science project, and introduces participants' attempts to expand the project's work to the national level. These efforts were crucial because project participants were faced with the fundamental question that confronts everyone once a significant period of funding or other support is over: how to continue the work they've started past the intense period of in-depth activity during which the initial reforms took place. What follows is a description of strategies designed to ensure the continuation of the specific activities initiated by the Women and Science project, most of which can be adapted to other circumstances.

## STEPS TOWARD INSTITUTIONALIZATION

Some of the smaller communities created through the Women and Science project made immediate plans to sustain the work already begun during the funded portion. At UW-River Falls, for example, Faculty Fellows wrote a proposal for a spin-off project at the site, a plan that eventually became the "trainer of trainers" model, in which DVPs pass their expertise on to Faculty Fellows, who then continue the project's work by serving as resident experts within their own departments and on their own and other campuses. This is an example of how changes that began during the project can not only be continued, but can also be implemented in other settings. In their role as trainers, the Faculty Fellows at UW-River Falls developed a workshop for science and math faculty at another campus, coordinated by an Advisory Board member there. Before the scheduled event, two of the organizers traveled to Superior to acquaint themselves with local science faculty and promote their interest in the workshop, which was then tailored to meet the needs of this specific audience. This preliminary visit, which included a meeting with the campus's Chancellor and Vice-Chancellor, mirrored the elaborate set-up and introductory activities Collaborative Community members had learned were necessary for a successful project. Because of the involvement of the campus's Women's Studies coordinator, an Advisory Board member and host for the workshop, the workshop and visits brought together faculty members who didn't normally meet, such as those in science and math and in Women's Studies and education. The workshop itself, which was later repeated with adaptations at UW-Oshkosh, included a number of hands-on exercises that placed faculty in the role of students, as well as providing them with a strong theoretical basis for those activities.

Another spin-off project from the Collaborative Community developed when Sherrie Nicol returned to UW-Platteville from her appointment as a DVP. During the following spring and fall, with joint funding from the NSF project and the Platteville campus, she organized a series of workshops presented by internal speakers and by visitors from campuses outside Wisconsin, targeting the College of Engineering. These speakers served as very short-term DVPs and presented material on course content, pedagogical issues, and classroom climate, visited the classrooms of the two Faculty Fellows and discussed teaching issues with them, and several also gave talks or colloquia for student organizations. While some of the early workshops got a low-key response, those events planted seeds of interest among the faculty and staff. Organizers learned from this experience that it was easier to build new connections and stimulate additional interest at other campuses because the Systemwide collaboration was already in place. A measure of the success of Nicol's project was the establishment of a permanent Women and Engineering Program with a full-time director, and the redesign of introductory courses in engineering, physics, and other science departments. Schools of engineering, which tend to have even fewer female faculty and students than other science departments, are in some ways the "final frontier" for women, and this campus is one of the few in the UW System that offers a full range of engineering courses. The UW-Platteville example, in which Nicol returned to her home campus with a plan to reform curriculum and faculty development, suggests how participation in a program like the UW's can not only revitalize the teaching of individual faculty members, but can lead to larger changes as well. Like the trainer of trainers model, this project also demonstrates the ripple effect that is key to the long-range success of the Women and Science Program.

From the inception of the project, and during the implementation of these many DVP activities, organizers recognized the importance of planning for the program's institutionalization. The Women and Science Program Advisory Board, which evolved over the course of the project, played a key role in this process, eventually including an administrator as well as a faculty member from each participating campus. This was also a way of keeping people across the state informed and stimulating interest in future projects among those who might not yet be actively involved. DVPs made presentations about their projects to the Board, Board members were invited to program retreats, and they were active in NSF site visits, meeting with evaluators and attending DVP workshops.

* * * * *

Before moving on to the ways in which this particular program has been institutionalized within the University of Wisconsin System, however, it is important to point out a specific type of curriculum reform that was initiated with the Women and Science project and that must be continued in order for the goals of curriculum reform and ongoing faculty development to be met. This is the important work of establishing new courses, an effort which is crucial to the integration of new content into the curriculum. In the following section, Faculty Fellow Kim Mogen describes the process by which a course on gender issues in science was developed and taught only once at UW-River Falls.

# Developing and Implementing a "Gender Issues in Science" Course

**Kim Mogen**

The idea for offering a course at UW-River Falls on women and science had been incubating in several faculty members' minds for some time. It wasn't until after the initial fall 1992 conference, however, that a committee started meeting to put together a syllabus. The 8-member committee, comprised of female and male faculty from the departments of Biology, Chemistry, Physics, Math and Computer Science, and Education, met throughout the spring 1993 semester. Using several syllabi obtained from similar courses taught at other universities as examples, they developed a course outline that addressed the women and science issue. The course, named WMST 220 "Gender Issues in Science," was added to the Women's Studies department's offerings.

The course had been approved by the college and university curriculum committees and was set to be offered for the first time during spring semester, 1995. No one stepped forward to teach it. While part of this reluctance was based on time and teaching load constraints, the other contributing factor, I believe, was a feeling of apprehension. After all, we had been trained and had focused our careers on learning and teaching the details of our respective sciences, not the social or historical aspects of science. Also, this course could not be a standard, familiar, lecture-based course. It would instead, by necessity, be driven by a considerable amount of student discussion of ideas. No wonder we, as traditional scientists, felt somewhat uncomfortable teaching the course.

However, Dr. Barbara Nielsen (Chemistry) and I (Biology) had served as

Faculty Fellows in the Women and Science Program since Fall 1992. Dr. Rhonda Scott-Ennis (Chemistry) had served on the committee which developed the Gender Issues and Science course and had been learning from our experiences. We decided to teach the class together, with myself as the lead instructor and them as regular contributors. The team teaching strategy was important as it overcame the problems of 1) not having enough time to research every topic oneself; and 2) apprehension that classroom discussions would collapse into silence.

Participation in the Women and Science Program had given us concepts to address in the course, which had not been articulated in the first syllabus. Dr. Cheryl Ney, still acting as the DVP on our campus, was extremely helpful and knowledgeable about specific methods and readings which might be useful in the course. As Cheryl and I modified the syllabus to reflect my new knowledge of women and science matters, we added a technology component, one laboratory activity, a compare/contrast teaching methods activity, and several ways of assessing student learning. Our course objectives were:

1. To introduce students to the nature of scientific knowledge.
2. To examine how women have participated in and contributed to the sciences.
3. To evaluate historical and current barriers women in science face and explore possible solutions.
4. To explore, both historically and currently, how science has been used to deny or discourage women from becoming scientists.
5. To discuss some of the feminist analyses of science and their implications.
6. To introduce students to electronic information/communication technology.

Cheryl Ney's prior experience with teaching gender and science topics to a mixed group of students was invaluable. We knew we had a very diverse group of sixteen students. Half were science/math majors, half were not (business, English, women's studies, social science); one quarter were male; one quarter were older, non-traditional students. We wanted the students to first think about what "science" is. Most of them, including the science majors, had experienced science only as a fact-filled body of data to be memorized. We tried to broaden their definition to allow that science is a way of observing and exploring the natural world and that all the "answers" will never be uncovered. That done, we chose the student projects with care. In order to get the non-science majors to believe that science does affect their world (and feel more connected to the class), we assigned project number one, "Who Does Science?". Project number two, "The Science/Gender Moment," was designed to get science and non-science majors thinking about how science and gender co-exist. In addition, both of these projects required short oral reports by the student within the first days of class. They had to talk to each other and start sharing thoughts and ideas! We purposely wanted to create, as quickly as possible, an open, friendly environment where the students could feel at ease to discuss ideas.

> "I liked the small class size because I wasn't so nervous about saying something stupid. The small class size also allowed for better discussions."

> "I liked that we could say what we really thought and not be ridiculed."

Anticipating that some students would still feel uncomfortable voicing their thoughts, we assigned journal entries for each lecture session (project #3). Here was a place they could communicate with the instructor, but in a more protected manner.

> "The journals stick out in my mind as being vital."

> "Disliked—yet is good communication."

Most current texts, by their womanless depiction of science, give the unvoiced impression that science is, and should be, a male-only profession. The purpose of Project #4 ("History of Women in Science") was to have the students discover that women have actually been doing science all along, perhaps under difficult situations, and in reduced numbers, but they have been doing it nonetheless, and not only are they capable of it, but they are good at it, too. Below are some quotes from students who responded to the question, "What did you find most valuable about this course?"

> "The history of women. I hadn't known there were so many out there!"

> "Getting information on the history of women in science and the discrimination women faced. I particularly like The Less Noble Sex because it helped me understand where some of this discrimination came from."

> "I was never aware of: 1) so many women in science's history, and b) how science, philosophy, etc. were used to define women in such a negative way. I felt both of these were of great importance to learn."

> "The lack of credit given to women throughout the history of science discovery. This would give me 'support' and a feeling of belonging if I were a (female) science major. I wish there was a diversity course in my major, like to learn about the barriers women face."

This topic did not appeal to all, however:

> "History is boring—keep it short."

For Project #5 ("Science Shaping Women"), the student were asked to focus on and explore a current example of how science has been used to define women. This should be the project in which they tie together much of what they have learned throughout the semester. It was the hardest project for them, probably because it required more critical thinking on their part and partly because we were entering the "end-of-the-semester burn-out." I would probably try to introduce the project earlier next time, or offer it as an alternative to the history project for those who are not turned on by history.

* * * * *

The development of a Gender and Science course illustrates the kind of curriculum development that can accompany a faculty development project that has as its goal the transformation of classroom practices, with attention to pedagogical, content, and climate issues. It also suggests directions for future curriculum and faculty development efforts if the goal of teaching science to all students is to be achieved.

## Future Directions

The preceding chapters have considered the multiple levels of development, reform, and collaboration on which the Women and Science Program has operated: among students and between faculty and students in the classroom; among faculty within their own departments; between faculty and administrators on individual campuses; and across departments, disciplines, colleges, and campuses throughout the UW System. Yet much more work needs to be done to sustain the collaborations that have started in this project and to build new ones, including those that will expand the work of reform to the national level. The ultimate goals of the Women and Science Program include ensuring that change is widespread and sustained, and if these aims are to be met, the changes in content, pedagogy, and climate that have already occurred must not only be continued, but developed and passed along to others in ways that are effective both in other contexts and cross-culturally.

Grants like the ones that enabled the original Women and Science project can galvanize faculty and administrators into making an impressive start on reform of the curriculum, classroom practices, and campus climate, and as participants have noted in previous chapters, can even push departments into dealing with controversies that have been simmering under the surface for a long time. In their interim report, evaluators Gloria Rogers and Judith Levy remarked on precisely this situation: "One of the Campus Coordinators reported that she felt there was a 'hidden goal' in that faculty could try new things under the sanction of the grant. It has empowered faculty to make change without criticism—provided a 'safe haven' for both women faculty and younger [male] faculty."[2]

Still, the long-term challenge lies in maintaining these changes so that the ways in which science is learned and taught are altered permanently, at both the local and the national level. Institutionalization needs to occur in a number of ways: change in how science is taught, including to whom it is taught; change in the practice of science itself (since many of the students will go on to become practicing scientists, as are the faculty members themselves); change in the place within the university of innovations like those promoted by the Women and Science Program; and ultimately, changes in the place of science in society. These are ambitious goals, none of which can be completely met through individual, short-term projects like this one, but such programs and their offshoots can provide the basis for long-term change.

In the rest of this chapter, program participants describe several of the challenges that remain after the initial funded portion of a reform project like this one is over—challenges that are especially important if the work of such programs is to expand to higher education in general and the wider society as well. This kind of broad national transformation both in the classroom (in pedagogy, climate, and content) and in faculty development activities has always been one of the basic concepts underlying the Women and Science Program, and a central feature of its goals and practices. All along, participants have given consideration to transferring what they have learned into settings that may be dramatically different from the large, statewide public university system described in this book so far.

Although the Women and Science Program began with an emphasis on diversity that included concern over the underrepresentation of ethnic and racial minorities in the sciences, for various reasons its focus came to rest finally on women, and both faculty and administrators recognize that a critical part of the reform agenda is thus still incomplete. The commitment to diversity is not merely a question of equity, or even of improving the situation for minority students and scientists. It is a matter of making knowledge more complete, enhancing science itself by incorporating the multiple perspectives of those who practice and are affected by it. Like all of the innovations and reforms described in this book, the inclusion of more, and more varied, points of view makes better science not only for members of underrepresented groups, but for majority students, teachers, and scientists as well, and this is particularly important as the project broadens to reach a national audience.

In Chapter 3, Catherine Middlecamp illustrated the importance of cross-cultural awareness in the classroom, the achievement of which depends heavily on the availability of faculty development activities. Systemic change, as the Women and Science Program has illustrated, becomes possible with a commitment to long-term efforts that focus faculty communities on scholarship in cross-cultural teaching. Efforts of this magnitude, and perhaps more, will be required to accomplish broad changes. The cross-cultural teaching Middlecamp considers refers to teaching students from different cultures, but the notion of "cross-cultural" can and should be expanded to include not only distinct ethnic and national cultures, but the cultures of specific locations as well, including the distinct cultures that are represented among different institutions.

* * * * *

Marc Goulet, a mathematician at UW-Eau Claire, offers a very specific example of how the methods and approaches learned in a project like this one can be applied in different pedagogical circumstances. His experience with a summer program for minority high school students makes it clear that the pedagogical lessons of the Women and Science Program can be transferred to contexts that involve students of ages, backgrounds, and demographic profiles that vary considerably from those involved in the UW project. As we've suggested throughout the previous chapters, faculty and administrators at post-secondary institutions that are quite unlike the UW System—smaller schools, private colleges and universities, and those with very different student populations—should take heart from his example.

# The Math and Science for Minority Students Program

**Marc R. Goulet**

In the summer of 1993 I began teaching in the Math and Science for Minority Students ([MS]2) Program at Phillips Academy in Andover, Massachusetts. This is a summer program which targets the African American, Hispanic/Latino, and Native American high school student population. From reservations and designated urban areas, rising sophomore high school students from these currently underrepresented groups in mathematics and science are selected to participate in the program for three consecutive summers. The students are chosen based on their demonstrated superior mathematical and scientific abilities, recommendations from their parents and teachers, and a face-to-face interview with the director of the (MS)2 Program. The students are provided travel to and from their homes to Andover, and full tuition, room, and board for the Phillips Academy Summer Session. Participants in the program live in dormitories with Summer Session students, but participate in an academic schedule which is specific to the program.

The academic component is rigorous and includes a three-summer sequence of courses in mathematics and science. The first summer consists of a biology course and a first-year mathematics course such as enriched algebra or precalculus. The second year consists of a chemistry course and a second-year mathematics course such as precalculus or discrete mathematics. The third year consists of a physics course and a third-year mathematics course such as discrete mathematics, dynamical systems, calculus, or a special topic. The program also includes an English composition component for the first two summers and a college counseling component for the third summer. The (MS)2 students take all of these courses exclusively with other (MS)2 students. The summer of 1997 marked the 21st Session of (MS)2, and my fifth year of involvement with the program.

## Women and Minorities in Mathematics

Since I was a graduate student teaching assistant, I have felt the need to participate in increasing the percentage of women and minority students who choose careers involving significant amounts of mathematics. I have taught in two summer Upward Bound programs at the University of Maine and at Oregon State University and also taught in the Educational Opportunities Program at Oregon State University. However, it was not until my involvement with the Women and Science Program in the University of Wisconsin System that I began to see clearly the role that pedagogy can have in achieving the goal of increasing the participation of underrepresented groups in mathematics and science.

The Women and Science Program has helped my development and growth as a faculty member through providing a network of people who are interested in looking hard at what we do in our classrooms and the impact that these teaching strategies have on our students. This network of people spans universities and disciplines and provides a rich support base. Buoyed by the Women and Science Program, I developed and structured courses in the (MS)2 Program to make more use of cooperative groups, reflective writing assignments, and an active laboratory component.

In our current Discrete Mathematics and Calculus Courses in the (MS)2 Program, we make extensive use of cooperative groups. The structure of the program serves as an excellent support network for the participating students, and the Phillips Academy environment serves as an excellent pre-college preparation. Living, working, and playing together for six weeks of the summer for three consecutive summers creates a comfortable environment for intellectual growth. Cooperation within the classroom nourishes this environment.

## The Classroom

In my classrooms in the (MS)2 Program, students belong to base groups of three or four students each, within a class of ten to twenty students. The week begins with an activity that the students engage in within their base groups. For the Discrete Mathematics and Calculus classes this takes place within a computer laboratory. The activity serves as the student's introduction to the material, and each group submits a laboratory report, which emphasizes the methods chosen to solve the problem, and records questions that remain unanswered in their minds. The next day, my teaching assistant and I lead a classroom discussion on the laboratory. We will pose questions to the class as a whole, and give between two and five minutes of conference time within the base groups for students to formulate responses. After this time, a rotating spokesperson for the group will share with the class their group's response. From time to time, I will give "expert" advice, but generally the students build their own knowledge from their experiences in the laboratory and class discussions. The students are then given a homework assigment and encouraged to work in their base group to complete it. Assessment is based on the activities, classroom participation, exercises, group and individual quizzes, essays, and a final examination.

The reaction of students, based on essays and course evaluations, has been generally positive. Students mention most often that they enjoy the group work, and the exhilaration of constructing their own techniques to solve problems. They complain mostly about the amount of work that they have to do in the course, and the fact that they are uncomfortable being asked to explore a concept in a laboratory without first being formally introduced to the ideas. As I am quite interested in choosing pedagogies which will serve to recruit and retain minorities and women in mathematics and science, the reaction of the students to this approach is very important. I have decided to try a slightly modified approach in the summer of 1997.

## The LAB Approach

The modification I have in mind responds to my own observations regarding the amount of time that students spend in reflective thought and their own acknowledged discomfort with the lack of formal lecture-style introductions to material. So, I will try what I have named the LAB approach: Launch, Activities, and Building understanding.

The base groups will remain in place, but now the launch component will set the context for the upcoming activity. It will include some lecturing, but will also include assigned readings from which students will formulate questions and begin the reflective learning cycle. The activity component will then serve to create experiences from which students can begin to glean answers to their earlier questions. The students will submit laboratory reports on their activities which include written responses to their earlier questions. In the next class period we will then begin to build understanding of the ideas through discussion. After this discussion, some groups will be encouraged to revise earlier written laboratory reports, to include their new understanding of the concepts. I am hopeful that this method goes some way towards alleviating the discomfort that the students have with adjusting to taking more responsibility for their own learning, and serves to support my efforts in encouraging the students to be reflective learners.

* * * * *

## Institutionalizing Change

In order for faculty to delve more deeply into curricular reform, it is usually necessary to ensure that they will be rewarded for that extremely difficult, time-consuming, and often controversial work. But it is also critical that reform be seen by both faculty and administrators as central to the mission of the university and, indeed, to higher education as a whole. Institutionalization not only helps to protect such projects from the hand-to-mouth struggle for funding, support staff, supplies, and other necessities that has plagued fields like Women's Studies. It also represents a concrete commitment on the part of administrators to the idea that the kinds of strategies developed by participants should be central to the way science is learned and taught. Whether at the departmental, campus, or national level, faculty and administrators must work together to accomplish a goal as large as transforming the way science is learned and taught.

It can also be crucial to seize the momentum created by a funded project like this one, and here again, administrators who take its goals seriously are key to its long-term success. In their interim report, evaluators Rogers and Levy quoted several administrators on the necessity of institutional support if the work of reform is to continue beyond the project's official end:

> "We're on a roll and we have to stay with it. . . . This program has created that stimulus." Deans commented that this activity "has to count [in the process of promotion, tenure, and review]." To make it work you "need administrative support at the unit level and a Dean who is willing to move money around to make this work." ("Interim Report," p. 15)

* * * * *

In establishing a permanent home for the Women and Science Program, administrators in the UW System have gone far beyond the level of support embodied in such efforts as the "trainer of trainers" plan and have demonstrated the kind of commitment to change that is needed if reform is to occur at the national level. In this section, Dean Michael Zimmerman of UW-Oshkosh, where the Women and Science Program is now housed, explains how it was established, and describes some of the specific plans that have been put in place to ensure its ongoing success. Both the challenges and the solutions he describes suggest a model other institutions might be able to follow.

# Ensuring the Future: Institutionalizing the Values of the Women and Science Program

**Michael Zimmerman**

When the Women and Science Program was originally founded it was created as a University of Wisconsin System-wide program funded by a grant from the National Science Foundation. While it is not an exaggeration to say that the program, in terms of the number of people reached, curriculum modified, and support offered, was more successful than anyone originally believed possible, as NSF support came to an end the program was in danger of being phased out. Because everyone involved was impressed by the good that had been accomplished in a relatively short period, this was a particularly disconcerting prospect. Discussions designed to find a way to institutionalize the program in the absence of continued outside financial support were given very high priority, but no obvious solution was immediately forthcoming.

The University of Wisconsin-Oshkosh, where I am dean, then stepped forward and proposed to house the program on its campus, providing salaries for a director and clerical support if all sister institutions would make modest contributions to help cover operating expenses. The proposal seemed to be a good one for all concerned. Faculty and administrators at UW-Oshkosh, looking to strengthen their commitment to gender-conscious pedagogies and to improve recruitment and retention statistics for women in the sciences, were convinced that the benefits of bringing the program to campus would more than offset the financial commitments. Individuals on the other campuses were pleased that the program would be able to continue its good work without imposing any fiscal hardship.

With all players throughout the University of Wisconsin System buying in to the UW-Oshkosh proposal, the program was placed on a secure footing for the foreseeable future and attention could be turned to the next major task: ensuring that the values and goals of the program were accepted broadly.

Needless to say, institutionalizing the values of the program is a much larger undertaking than that of institutionalizing the program itself. While it is relatively easy to speak generically and somewhat glibly about the importance of the Women and Science Program, it requires a much deeper commitment to make significant alterations in pedagogical style, in material presented, and in the way the roles of students and instructors are conceptualized. But these are the very things that are at the heart of the program. And these are the very things that must be accomplished if the program is to be considered successful in the long run.

As we look to the future, and as we build support for the goals of the program, there are two sets of issues that have to be resolved, both dealing with the composition of the program's constituencies. The first issue is whether the program should focus on students or faculty, while the second is whether the program should concentrate its efforts within the state of Wisconsin or whether it should broaden its horizons and attempt to influence the way university-level science instruction is performed across the nation. After considerable discussion, program participants have come to the optimistic conclusion that the program has reached the stage of maturity that would allow it to successfully deal with all of these constituencies.

An ambitious agenda of this sort is critical because of the nature of the program's goals: systemic, long-lasting, and meaningful change in science pedagogy can only occur if an integrated approach is taken to the topic. A simplistic example will demonstrate the necessity of an integrated approach. If, for example, the tools for curricular reform are provided to interested junior faculty members in a chemistry department but senior colleagues are skeptical about, and perhaps even insulted by, the new methodologies proposed, it is unlikely that even the most committed of junior faculty members will be able to successfully implement change. Similarly, without an appropriate introduction to the goals of the program and without an appropriate support structure in place, students used to the standard classroom environment might well react negatively at first to alternative learning strategies. Administrators and faculty members serving on tenure, reappointment, and promotion committees must be educated about these problems and must not overreact to teaching evaluations filled out by students experiencing innovative classroom environments.

In its attempt to deal with just these sorts of problems, the Women and Science Program has decided to focus its attention on all aspects of curricular reform. The call for proposals for inclusion in the program's annual curriculum reform Summer Institute, for example, explicitly states that campus teams must include both tenured and untenured faculty members. Additionally, the participation of department chairs, course coordinators, or other administrators, while not required, is strongly recommended. Towards this same goal, the program has made a portion of its limited operating budget available for University of Wisconsin teams to hire student assistants to help with the testing and implementation of the new curriculum. The intent of all of these strategies is to build as large a base of support for curricular change as possible.

Structuring the Institute in this fashion helps to achieve yet another critical goal of the program: building a community of professionals committed to curricular reform. It is imperative that faculty and staff members struggling to bring innovations into their classes, sometimes in the face of opposition by both colleagues and students, know that they are not alone in their efforts. Being able to communicate with others who have been or who are going through the same types of experience is absolutely essential for success. Because of the program's belief that the broader the consultation possible, the more likely that meaningful support will be offered, attempts have been made to build on-campus, Systemwide, and national communities. Indeed, this desire was one of the factors that led to the decision to expand the work of the program to the national level.

Communities are not easy entities to construct, however. To maximize the probability of success, the program has decided to move forward in a number of directions at one time, five of which are worth mentioning. First, an electronic bulletin board has been established for discussion of ideas and problems. As with many bulletin boards, it is frequently difficult to generate interesting threads. Nonetheless, even occasional discussions serve a valuable purpose.

Second, the program helps make arrangements for intercampus visits and presentations. Such visits were the model for the national consulting service established by the program. It is axiomatic that anyone serving as an academic consultant learns something from an on-campus visit that can be brought back to the home campus, and we have found that in addition to sharing their expertise with others, program consultants do bring back new and exciting ideas from the campuses they visit.

Third, the program sponsors a yearly retreat at which ideas and teaching strategies are discussed. As with any conference, one of the most important aspects of the retreat is the informal opportunities colleagues have to interact with one another. By broadening the retreat to participants in the annual curricular reform institute, the hope is that a wider array of pedagogical strategies will be presented and that interpersonal connections made over the summer will be reaffirmed and strengthened. Just this past summer, for example, a number of pairs of institutions began discussing the possibility of undertaking some collaborative, interactive classroom experiences through distance education technology. How and if these ideas take shape remains to be seen, but it is exciting that such discussions are at least taking place.

Fourth, beginning with faculty members newly hired to teach during the 1998-1999 academic year, the program now offers a System-wide orientation program. Departmental mentors as well as the new faculty members are invited to participate. In addition to structuring a discussion of alternative pedagogies for professionals very early in their teaching careers, this workshop brings together people with similar interests. The hope is that the kind of broad-based community not possible on any single campus will result.

Finally, the program now produces a newsletter. By reporting on new initiatives and commenting on past successes, the newsletter helps demonstrate that the people who comprise the Women and Science Program are active and vibrant. The newsletter is distributed to campus administrators across the System to ensure that they are fully informed of ongoing activities.

A sense of community is every bit as important for students as it is for faculty and staff. For that reason, the program has been supportive of campuses that are in the process of replicating the successful Women in Science and Engineering Residence Hall begun on the Madison campus. It is important for competent, capable women to realize that they can succeed in one of the sciences and that the desire to do so is not something that should be sublimated or something that is looked on as an embarrassment.

The long-term success of the Women and Science Program is thus entirely dependent on the creation and maintenance of a community of professionals who support one another in their work to improve the quality of undergraduate instruction. In addition to providing the opportunity for these people to interact with one another, the Women and Science Program is taking as many steps as possible to ensure that ample communication is taking place and that there is a sense of excitement about programmatic activities. If the program is capable of maintaining communication and building excitement, the probability is that the program will continue to grow and thrive well into the future.

* * * * *

In this section, Zimmerman has outlined the kinds of activities and organizational structures that need to be undertaken in order to ensure institutionalization and extend the collaborations in Wisconsin to a national level. Significantly, he suggests that this institutionalization comes not merely with more activities and structures, but also through institutionalizing the values of the program, transmitted—through the communities that have been built already—to new participants and their communities. Below, the first permanent Director of the Women and Science Program addresses in more detail just how a project that was initially externally funded moves towards institutionalization. She provides a description of the kinds of activities and structures that followed the initial project. Importantly, these activities offer the kind of opportunities that are needed for the cultivation of the values of the project, as well as their further development and extension to a much wider audience.

## The Program Director's Perspective

In writing about the year of transition between the NSF-funded project and the permanent Women and Science Program, Heidi Fencl, the program's current Director, repeats a point on which participants, administrators, and evaluators have been unanimous: "The challenge . . . was to keep the momentum going: to maintain a strong sense of community among the participants, to interest more and more UW System educators in gender-conscious teaching, and to communicate information from the [project] to national audiences. These will continue to be areas of emphasis of the program in the coming years."

Fencl emphasizes, too, the fact that the program must fulfill local needs within the UW System:

> Many hundreds of faculty and staff are required to teach all the science and mathematics courses offered at UW System institutions. The ultimate goal of facilitating systemic change in the way science, mathematics, and engineering courses are taught requires much more effort within the UW System itself. For this reason, a priority of the Women and Science Program is to offer System-wide workshops for incoming faculty and academic staff in mathematics and science.
>
> The first of these annual workshops took place in October, 1998. All incoming mathematics and science educators, as well as some senior mentors, are invited to attend a two-day workshop held during each fall semester. The workshops focus on student-centered teaching, and have a practical emphasis so that information presented is immediately applicable as participants design and revise their teaching materials. Funding was also obtained to offer summer support to a subset of the educators attending the workshops. Support is awarded on a competitive basis to participants who show a strong commitment to incorporating gender-conscious teaching strategies into one of their assigned courses or laboratory sections.
>
> Response to the workshops has been extremely positive. Participants report in follow-up surveys that they incorporate strategies presented at the workshop into their teaching. Equally important, they report that they are talking about the workshop with departmental colleagues and are exchanging ideas about ways to improve their teaching.
>
> By continuing the workshop series, it is hoped that most incoming and many senior science and mathematics faculty members will have a chance to experience cooperative learning, open-ended laboratories, problem-centered learning,

> and other student-oriented pedagogies, and see that they can, and should, play an important role in learning and doing science.
>
> There is an additional need that, over time, new faculty workshops will help to fill: to continue to increase the types of gender and teaching issues that are addressed by the Women and Science Program. As this book describes, the history of the project is centered on faculty development and course reform. It is no surprise that many educators are interested first in better teaching, and become aware of gender and climate concerns through those avenues. However, it is also true that once teachers are aware of studies on gender and science, they are more conscious of factors inside and outside of their classrooms that impact their female students. By introducing a large number of faculty members to gender-conscious teaching, it is likely that departmental, as well as classroom, climates will become warmer.

Finally, the establishment of a permanent Women and Science Program on the UW-Oshkosh campus allows participants to continue their commitment to being "trainers of trainers" beyond the UW System. Fencl writes:

> Another exciting project underway since 1997 is the UW System Women and Science "Science, Gender and Community" Curriculum Reform Institute. The Division of Undergraduate Education of the National Science Foundation provided funding to establish this Institute, which offers educators throughout the nation an opportunity to learn about gender-conscious pedagogies, and to incorporate them into their own courses. Because of the continuing need to reach more faculty members in the UW System, funding was also obtained to expand the workshop associated with the Institute to include additional teams of educators from within the UW System.
>
> The first annual Institute workshop was held in June of 1997. During the week-long event, 16 teams composed of faculty members, administrators, or other educators from a single institution came together to work with mentors on course development or reform projects. The leaders included past Distinguished Visiting Professors (DVPs) and Faculty Fellows from the original project as well as other national leaders in gender-conscious education. Teams came from California, Washington, Wisconsin, Illinois, Texas, Louisiana, Maryland, New York, Rhode Island, and the District of Columbia. The projects that they developed were as diverse as the participants. Some teams established outlines, teaching approaches, and activities for Women and Science Courses; others developed interdisciplinary general education courses. Several groups worked on computer science courses and transition-to-work programs; and traditional introductory courses in biology, chemistry, and physics were reworked to be more gender-, and student-, conscious. These projects have been implemented on teams' home campuses where they directly impact thousands of students annually. Outlines and other materials from this and subsequent Institutes are available to wider audiences on the World Wide Web at http://www.uwosh.edu/wis/.

The design of the Curriculum Reform Institute is also intended to foster growth within the UW System Women and Science Community. The Institute does not end with the summer workshop. Workshop participants and leaders continue to interact through media such as conference listservs, and all are invited back to Wisconsin to attend the Women and Science Spring Retreat the following May. The retreat is a chance not only to renew contacts and connections, but also to present updates on new and continuing projects. It is hoped that many Institute participants will establish close ties to the UW System Program, and, in fact, many participants at early Institutes now serve as Institute mentors and in other leadership roles.

A related component of the Institute is a Women and Science Consulting Service, which matches interested campuses or departments with leaders from the program. The leaders make site visits to the campus, and are able to address a wide variety of situations. The visits are flexible in format and length, and are arranged according to the needs of the contracting campus. Topics can include gender-conscious pedagogies, assessment, including the contributions of women in content courses, or other areas.

During the first years following institutionalization, energies of the Women and Science office have gone to developing System-wide faculty development programs. Large projects such as the opening workshops and Curriculum Reform Institute complement smaller events such as special topics workshops, program retreats, distance education discussion sessions, and campus visits. With this core of centrally coordinated workshops and activities, creativity now can be directed towards supporting individuals on each campus as they develop and implement their own local Women and Science events.

There are many other projects ahead for the UW System Women and Science Program. Further evaluations of attitudes and also of retention rates will need to be considered, and the program needs to continue to expand its sphere of influence. However, the past years have shown that DVPs, Faculty Fellows, and other educators who became part of the program under the original NSF-funded project are truly committed to the institutionalized program, and that recent activities, too, are producing dedicated leaders for future events. The answers to the question "What should we do next?" are many and varied, as are the people willing to see the projects through. With such a community of leaders, the future looks very exciting.

## NOTES

1. Barbara Brownstein, "Site Visit Report, April 14-16, 1994," p. 5.

2. Judith Levy and Gloria Rogers, "Interim Evaluation Report," Spring 1993-Spring 1994, p. 13; subsequent quotations cited in text as "Interim Report."

# Epilogue

Jacqueline Ross

> With increasing numbers of women going into graduate study in science, wherever you have a critical mass of women . . . you also have a feminist presence.
>
> —Ruth Bleier

As I hope we have demonstrated in this book, the collaborations developed in the course of the Women and Science Program have continued to flicker, creating series of new linkages and models for educating women in the sciences. Inspired by the ideas and accomplishments of Ruth Bleier, we have sought to translate her vision through systemic changes in the education of scientists, with special attention to women scientists, for the future. The first phase of this project, with its Distinguished Visiting Professors and host communities in the University of Wisconsin System, is over. As Michael Zimmerman and Heidi Fencl have explained, the new phase has begun, and there is much to be done.

This book also chronicles, through the lens of our program, the process of passing on the tradition of women and science. In the spring of 1997, the late Ethel Sloane, our first Distinguished Visiting Professor, was honored upon her well-deserved retirement after 37 years of service at UW-Milwaukee and in the UW System. Yet her contributions to the program, including her special approaches to the biology of women, continue through the ongoing work of her Faculty Fellows and others.

The other DVPs have returned to their home campuses, continuing to publish and teach about issues relating to feminist approaches to science. Several of them, along with many other program participants cited in the preceding chapters, have taken part in the annual Women and Science Curriculum Reform Institute, faciliating new collaborations with science faculty from around the country. Earl Peace is administering a major NSF-funded curricular reform project in the UW-Madison Chemistry Department, while continuing his valuable contributions as an active member of the Women and Science community. Rebecca Armstrong, the former director of our program, has left to complete her Ph.D. and pursue a new career, and has passed the torch to Heidi Fencl to carry out our goals in new and expanded directions.

Looking back, it seems that Cheryl Ney and I have been carrying on discussions for several years on all sorts of issues relating to feminism and science. Working with Laura Stempel on this book, an evolutionary process of its own, has marked another fruitful and enjoyable stage in our collaborations. As we move on in our own directions, we hope that *Flickering Clusters* will encourage our readers to explore variations on the approaches to learning and teaching described here, and to participate in the extended Women and Science collaborative community.

# Appendix A

## Evaluators' Report

# "Science, Diversity and Community: Revitalizing Introductory Curricula"

**Dr. Judith Levy and Dr. Gloria Rogers**

**Final Report**
**May 1994 – May 1996**

This report represents the second and final report on the assessment of the Women and Science program of the University of Wisconsin System Women's Studies Consortium. This report describes progress made by the project in meeting its goals and in implementing improvements; it also discusses the implications of the project outcomes with respect to the institutionalization of the project at the University of Wisconsin, and the efficacy of the project model for institutional change.

This report includes a summary of the design, methodology and results of the assessment of the Women and Science Project. It also presents results from the evaluation of the activities of the last three of the seven Distinguished Visiting Professors (see Table 1), and cumulative data on Faculty Fellows. The evaluations of the first four Distinguished Visiting Professors listed in Table I were presented in the Interim Report on the first year of the project (April 1993 to April 1994). The Interim Report, which is presented as an appendix, includes preliminary analysis of student questionnaire results from the Collaborative Community and excerpts from a variety of focus group sessions during the first year of the project. These two reports together provide an overview of this complex, multi-year project.

**Table 1: Distinguished Visiting Professor**

| Name | Term | Field | Home Institution | Host Institution |
|---|---|---|---|---|
| **Reported in Interim Report:** | | | | |
| Ethel Sloane | Spring 1993 | Biology | UW-Milwaukee | UW-Waukesha |
| Sue V. Rosser | Fall 1993 | Biology Women's Studies | U of South Carolina-Columbia | variety |
| Sherrie Nicol | Fall 1993 | Mathematics | UW-Platteville | UW-Eau Claire* |
| Cheryl Ney | Spring 1994 Fall 1994 | Chemistry | Capital University | UW-River Falls |
| **Included in This Final Report:** | | | | |
| Vera Kolb | Fall 1994 | Chemistry | UW-Parkside | UW-Madison |
| Danielle Bernstein | Fall 1995 | Computer Science | Kean College | UW-Stevens Point |
| Judith Heady | Fall 1995 | Biology | UM-Dearbom | UW-La-Crosse |

*The UW campuses at Eau Claire, River Falls, and Stout were referred to as the Collaborative Community.

## Assessment Design

The goals of the project evaluation were both summative and formative: to determine the effectiveness of the project in meeting its goals and to provide periodic feedback on the performance of the project for continuous improvement and fine-tuning of the project during its implementation stage. The evaluation scheme was designed to complement the complex project design in which Distinguished Visiting Professors (see Table 1), who had successfully implemented teaching innovations at their home institutions, taught students in introductory courses at the host institution, and worked with designated Faculty Fellows at the host institution to develop courses with improved climate, content and/or pedagogy.

Project characteristics of special significance to the design of the evaluation included the implementation of the project on multiple campuses, in multiple scientific disciplines, and involving multiple study groups (Distinguished Visiting Professors, Faculty Fellows, and students). The anticipated amplification effect, whereby each Faculty Fellow who worked with a Distinguished Visiting Professor would in turn teach an introductory course with innovative features, affected the timeline of the assessment strategy. Further complexity of design resulted from the fact that each Distinguished Visiting Professor essentially conducted a mini-project with a unique initiation point and specific combination of innovations of climate, content, and pedagogy. As the project evolved from one focusing on curriculum development goals to one with an emphasis on faculty development goals (such as the acquisition of knowledge and understanding of the scholarship on gender studies, and training of faculty to facilitate the institutionalization of the project), the evaluation plan also evolved.

## Assessment Methodology

The assessment methodology included periodic quantitative and qualitative data collection of students, Distinguished Visiting Professors, Faculty Fellows, other faculty members, and campus administrators. Assessment of students in sections taught by Distinguished Visiting Professors and in other non-targeted sections of the same course consisted of focus groups and pre- and post-course attitudinal questionnaires. The evaluators interviewed the Distinguished Visiting Professors and observed them in their classes. The methodology for studying the Faculty Fellows consisted of focus group sessions and survey instruments to assess changes in their knowledge and use of feminist pedagogies as well as to determine the effectiveness of various Distinguished Visiting Professor activities in bringing about these changes. As the project's emphasis on faculty development increased, additional survey instruments were developed that also explored factors that facilitated continued implementation of project goals by Faculty Fellows after their participation in the project ended. Other faculty members in participating institutions were also involved in focus group sessions to assess the influence of the project on the attitudes and knowledge of faculty members who were not Faculty Fellows. The evaluators also interviewed administrators on participating campuses, such as Department Chairs, Deans, and Vice-Chancellors, to learn more about the climate for change in the targeted departments and on these campuses.

## Formative Evaluation and Project Refinements

One of the goals of the evaluation scheme was to provide periodic feedback to the project administrators so that improvements could be made during the project implementation phase. This goal was effectively accomplished; feedback was provided to project administrators at the end of each site visit by the evaluators, and the project administrators were responsive in fine-tuning the project. Although fully recognizing that what works for one institutional culture may not always work for others, the evaluators present here a review of some of the project refinements that seem broadly applicable, such as changes in administrative organization and communication, and refinement of project goals and design.

***Communication.*** Early in the project when the administrative organization was being developed simultaneously with the initial implementation of Distinguished Visiting Professors at some sites, focus groups sessions with participants on host campuses raised issues of administrative organization and communication. Of special concern was communication about expectations and rewards for faculty who participated in the project and about what was expected of participants and the clarity of timelines and deadlines. The project administrators responded by implementing introductory meetings on host campuses for all participants and campus administrators at which expectations and deadlines were addressed. In addition, and possibly more importantly, the goals and design of the project were clearly articulated and a more collaborative relationship between project administrators and participating faculty members was established.

Another related communication issue was the perceived lack of sufficient lead time given for invitations for proposals from departments and campuses for Distinguished Visiting Professors. Faculty acceptance of and participation in project activities was facilitated by widespread faculty involvement in the definition of the campus proposals for project participation. This required sufficient lead-time for faculty involvement. Therefore, modifications were made which provided early, ongoing, and clear communication between project administrators and faculty members. Faculty Fellows and Distinguished Visiting Professors saw this communication as critical to the success of the project.

***Faculty development.*** Periodic evaluation and feedback led to recognition that the long term effectiveness of the project model was dependent on faculty development in areas of gender studies, curriculum and pedagogy. As a result, there was increased project emphasis on building cadres of faculty who could help, support and challenge each other to continue to pursue the project goals. Faculty Fellows were trained to conduct workshops for other faculty and share what they had learned from Distinguished Visiting Professors about gender studies and pedagogy.

The observation that the changes being implemented were characterized by their incremental nature and not by their discontinuity led to recommendations that Distinguished Visiting Professors and Faculty Fellows participate in the project for terms of one academic year rather than one semester. A modification of the model was tried where one of the Distinguished Visiting Professors, Cheryl Ney, stayed at her host campus for an additional year and helped to train her Faculty Fellows to themselves become trainers of other faculty members and share their new knowledge and skills. The comfort level of these Faculty Fellows with the role of "trainer" was greater than that for the Faculty Fellows who worked with their Distinguished Visiting Professors for only one semester. This was attributed to the fact that Cheryl Ney's Faculty Fellows had at least one semester where they observed and attended workshops before they began training others. On the campuses where the Distinguished Visiting Professor was only in residence for one semester, the Faculty Fellows were expected to become trainers at the same time they were being introduced to the project and without the benefit of having time to observe and reflect on their experiences. As the Faculty Fellows had not had an opportunity to implement the new pedagogy in their own classrooms, this expectation was viewed by the Faculty Fellows as being too demanding and ineffective. This observation supports the idea of longer terms of interaction for Distinguished Visiting Professors in this project model.

## The Collaborative Community as a Case Study

The other assessment plan goal was to determine the effectiveness of the project in meeting its goals. The evaluators have earlier presented student (see addendum to Interim Report 1) and faculty data (see attached draft of monograph chapter) for the "Collaborative Community" of the University of Wisconsin System campuses (at Eau Claire, River Falls, and Stout) that support the positive impact of the innovative pedagogy of the Distinguished Visiting Professors on student confidence and attitude about introductory science classes, and the effectiveness of this mentoring model in promoting faculty professional and personal development. The Collaborative Community is unique in several ways. Although the diversity of types and sizes of campuses participating in this project do not permit aggregation of all student data from all campuses into one data set, the number and homogeneity of campuses in this collaboration and the period of data collection yielded a large data set on students as well as faculty data over time. In addition, only at the Collaborative Community were the Faculty Fellows able to interact with their Distinguished Visiting Professor for more than a single semester, thus magnifying the effectiveness of the project. These unique aspects made it easier to observe the very positive impact of the project methodology on student and faculty participants.

## Final Three Sites

The evaluators present here the student data for three University of Wisconsin campuses — UW-Madison, UW-Stevens Point, and UW-La Crosse.

## Students

In the Interim Evaluation Report, data were reported which compared student responses to a questionnaire given at the end of the semester to a questionnaire given at the beginning of the semester. Results were analyzed comparing the responses of students in classes taught by the Distinguished Visiting Professor,

Faculty Fellows and, when possible, other faculty who were teaching the same course. The questionnaire was designed to gather data on student demographics and students' self-report on their level of confidence in their ability to *understand* the subject area (i.e., chemistry, biology, computer science) and their *attitude towards* the subject area. Students were asked to reflect on what their confidence and attitude was at the beginning of the term and what their confidence and attitude was *as a result of the class*. The Interim Evaluation Report illustrated only the data from the Collaborative Community (UW-Eau Claire and UW-River Falls). The following illustrates the results from UW-Madison, UW-La Crosse, and UW-River Falls. These sites participated in the project from Fall 1994 through Fall 1995.

## UW-Madison

There were 1643 surveys completed on the Introductory Chemistry class on the UW-Madison campus. For the Distinguished Visiting Professor's section, there were 234 surveys returned, Faculty Fellows (2) had 359 returned, and for other faculty (4) there were 1049 surveys returned. The student populations were very homogeneous. Tables of respondents by category are shown below:

**Sex**

| | |
|---|---|
| Females | 842 |
| Males | 801 |

**Classification**

| | |
|---|---|
| Freshman | 1357 |
| Sophomore | 163 |
| Junior | 72 |
| Senior | 42 |
| Other | 17 |

**Race**

| | |
|---|---|
| African American | 28 |
| American Indian | 10 |
| Asian/Pacific Islander | 101 |
| South East Asian | 27 |
| Caucasian | 1402 |
| Latino | 26 |
| Other | 48 |

**Course Requirement**

| | |
|---|---|
| Required | 1493 |
| Elective | 143 |

**Age**

| | |
|---|---|
| Under 22 | 1568 |
| 23-30 | 70 |
| 31-40 | 10 |
| 41-50 | 3 |

Students were asked to reflect upon a measure of their confidence in their ability and their attitude toward the subject matter both prior to taking the course and *as a result of taking the course*. The items were on a Likert-type scale with "I" being "Low" and "9" being "High." The questionnaire required the students to reflect on their level of confidence/attitude prior to taking the class. These ratings were subtracted from their post-course ratings on the same variables. This new variable is reported as a "mean difference" for each group (female/male) and represents their perceived "growth" for each variable. For the UW-Madison campus, the mean difference scores for students' **level of confidence** are reported below in Figure 1.

**Figure 1. Mean Differences in Pre- and Post-Course Confidence Levels of Students: UW-Madison Campus**

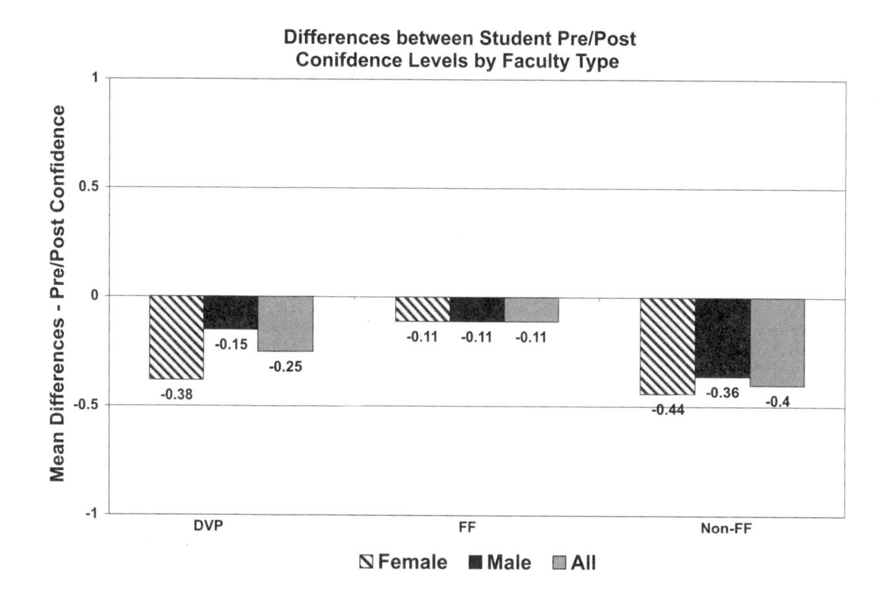


The results shown in Figure I indicate that, as a group, students in all sections were more confident in their abilities in Chemistry at the beginning of the course than at the end of the course. In fact, when looking at students' mean differences by individual faculty members, there was one FF (+. 1 3) and one non-FF (+. 004) whose students reported positive gains (even though they were very small).

A similar result was found on the item that asked students to indicate their **attitude towards** Chemistry at the beginning and the end of the semester. The mean differences between the two ratings are shown in Figure 2.

**Figure 2. Mean Differences in Pre- and Post-Course Attitude Levels of Students: UW-Madison Campus**

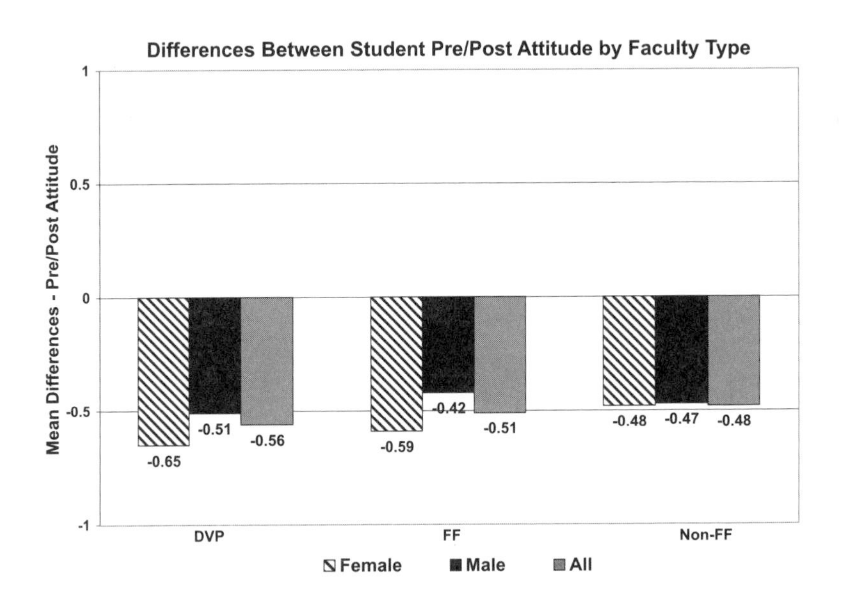


The results shown in Figure 2 indicate that, as a group, students in all sections reported a more positive attitude before taking the Chemistry course than at the end of the course. Unlike the confidence scores, this finding held for all sections regardless of faculty member. There were no statistically significant differences between the mean difference scores of female and male students nor among faculty groups. However, there were some statistically significant mean differences of student scores between the individual faculty members.

## UW-STEVENS POINT

There were 40 surveys completed in two classes of Computer Science at UW-Stevens Point. The student populations for the two classes were very homogeneous. Demographic data reported are given below.

**Sex**

| | |
|---|---|
| Females | 11 |
| Males | 28 |

**Classification**

| | |
|---|---|
| Freshman | 17 |
| Sophomore | 9 |
| Junior | 7 |
| Senior | 4 |
| Other | 1 |

**Race**

| | |
|---|---|
| African American | 0 |
| American Indian | 0 |
| Asian/Pacific Islander | 1 |
| South East Asian | 1 |
| Caucasian | 37 |
| Latino | 1 |
| Other | 0 |

**Course Requirement**

| | |
|---|---|
| Required | 26 |
| Elective | 14 |

**Age**

| | |
|---|---|
| Under 22 | 31 |
| 23-30 | 6 |
| 31-40 | 2 |
| 41-50 | 1 |

Figure 3 illustrates the mean differences for students' pre- and post-course level of confidence in their ability to understand computer science. There were no statistically significant differences between the mean differences of female and male students nor Distinguished Visiting Professor and Faculty Fellows. In all cases, students' level of confidence improved from the beginning to the end of the course in computer science.

**Figure 3. Mean Differences in Pre- and Post-Course Confidence Levels of Students: UW-Stevens Point**

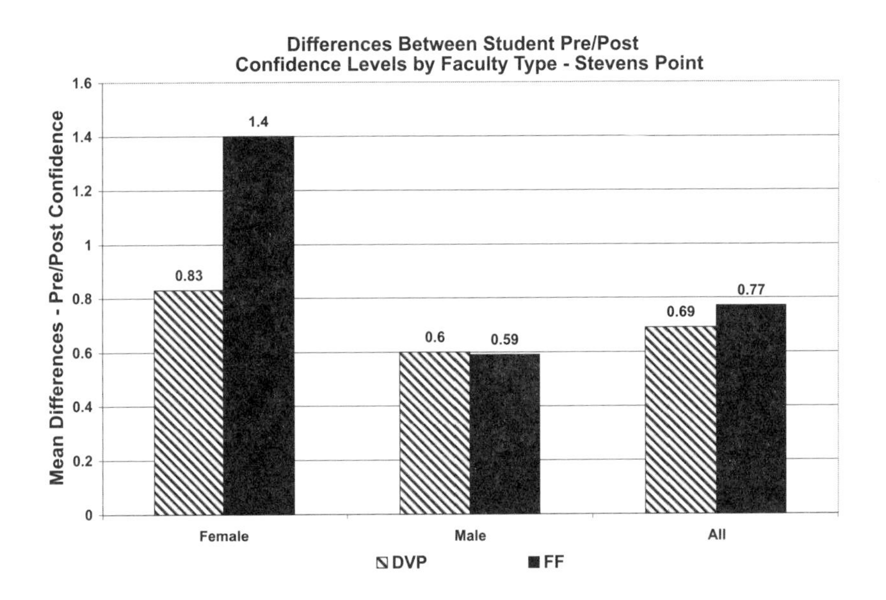


The students' mean pre- and post course score differences in their **attitude towards** computer science is illustrated in Figure 4. Because the data sets are small, the differences noted are not statistically different for any groups (faculty or students by sex). For example, in the DVP's class there were only 6 female students as compared to 10 male students. The FF had 5 female students and 17 male students.

**Figure 4. Mean Differences in Pre- and Post Course Attitude Levels of Students: UW-Stevens Point**

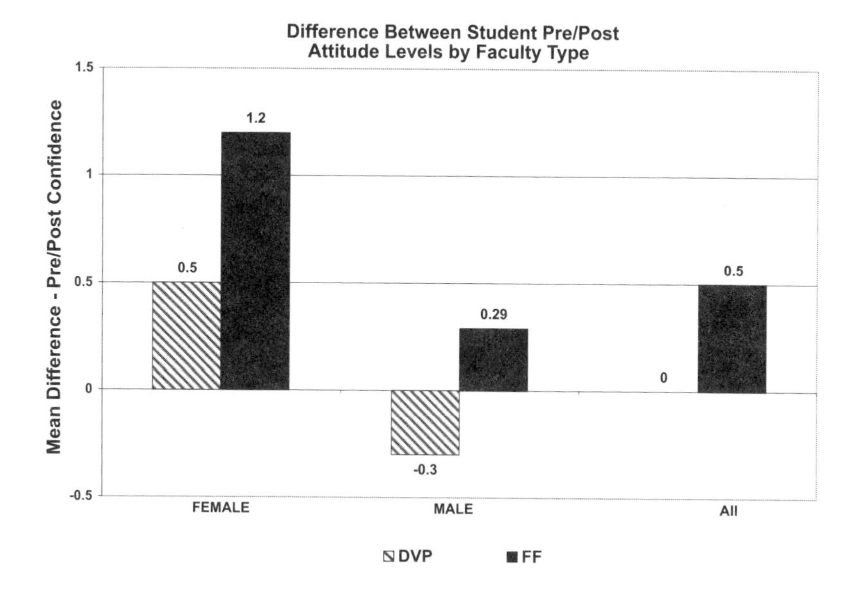

## UW-La Crosse

There were 135 student respondents in the Introductory Biology course at UW-La Crosse. Although women are disproportionately represented, generally the student demographics are homogeneous. The data are given below:

**Sex**

| | |
|---|---|
| Females | 81 |
| Males | 54 |

**Classification**

| | |
|---|---|
| Freshman | 104 |
| Sophomore | 25 |
| Junior | 3 |
| Senior | 3 |
| Other | 0 |

**Race**

| | |
|---|---|
| African American | 0 |
| American Indian | 3 |
| Asian/Pacific Islander | 1 |
| South East Asian | 1 |
| Caucasian | 125 |
| Latino | 4 |
| Other | 0 |

**Course Requirement**

| | |
|---|---|
| Required | 127 |
| Elective | 7 |

**Age**

| | |
|---|---|
| Under 22 | 127 |
| 23-30 | 5 |
| 31-40 | 1 |
| 41-50 | 2 |

**Figure 5. Mean Differences in Pre- and Post-Course Confidence Levels by Students: UW-La Crosse**

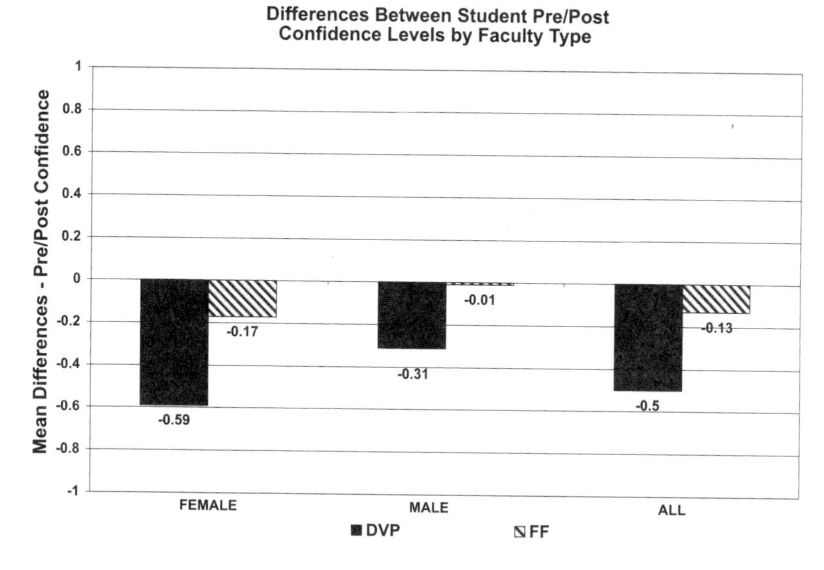


The pre- and post-course data indicate that, as a group, for females and males in both sections there was a decrease in their **level of confidence** in their ability to understand biology. The differences are illustrated in Figure 5. There were no statistically significant differences between or within the groups.

Similar findings are reported for the mean differences in students' reported **attitude towards** biology. Figure 6 illustrates the differences for their pre- and post-course ratings. It is important to remind the reader that all of the differences reported are based on a nine-point rating scale.

**Figure 6. Mean Differences in Pre- and Post-Course Confidence Levels of Students: UW-La Crosse**

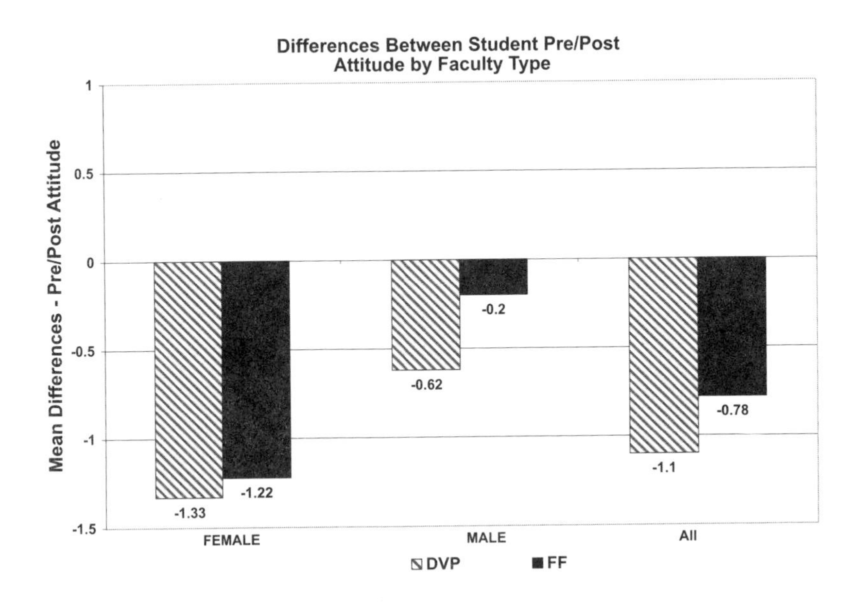


## FACULTY

The increasing emphasis on faculty development during the project made the collection of data on faculty a critical piece of the assessment. Faculty Fellows were surveyed to gather data on their experience in the project. Questions were asked to determine the effect of the project on their knowledge of issues related to instruction in science and/or mathematics and their familiarity with feminist pedagogical techniques. Faculty Fellows were also asked to provide general information about their experience in the project and the project in general. There were a total of 13 Faculty Fellows who responded (8 female, 5 male) to the May 1995 survey. As a group their responses about the project were positive. Because the number of respondents is small only descriptive statistics are reported here. The table below illustrates the responses of the faculty on selected items from the survey.

### Table 2. Faculty gains in knowledge of classroom techniques

| The items below were rated on a seven point scale with 1=Unfamiliar and 7=Very Familiar | All | Female | Male |
|---|---|---|---|
| **Prior to my participation** in (this project) my familiarity with teaching/learning pedagogy was: | 3.9 | 3.6 | 4.4 |
| **As a result of my participation** in (this project) my familiarity with teaching/learning pedagogy is: | 5.7 | 5.6 | 5.8 |
| **Prior to my participation** in (this project) my familiarity with feminist pedagogical techniques .... was: | 2.6 | 2.5 | 2.8 |
| **As a result of my participation** in (this project) my familiarity with feminist pedagogical techniques .... is: | 4.8 | 5.1 | 4.4 |

For both men and women faculty respondents, there were perceived gains in what they had learned about teaching and learning strategies and feminine pedagogy. This quantitative data supports information from the focus groups with faculty at the various sites of the project. The survey also asked faculty to respond to a number of items to gain information on their experience, and the future of the project on their campus. The table below illustrates some selected items.

### Table 3. Faculty responses on general items.

| The items below were rated on a seven point scale with 1=Strongly Disagree and 7= Strongly Agree | All | Female | Male |
|---|---|---|---|
| As a result of this project, I have changed what I do in the classroom and/or laboratory. | 5.7 | 5.9 | 5.4 |
| As a result of this project, I have a greater understanding of students. | 5.2 | 5.5 | 4.8 |
| As a result of this project, I have a more positive view of myself as an individual. | 5 | 5.4 | 4.4 |
| As a result of this project, I have a more positive view of myself as a professional. | 5 | 5.6 | 4.6 |
| 1 will continue to use/implement the goals of the project after the NSF grant expires. | 5.6 | 5.5 | 5.8 |
| My participation in this project has been encouraged by my department. | 3.5 | 3.3 | 3.8 |
| It is likely that my **department** will encourage the implementation of goals ... after NSF grant expires. | 3.9 | 3.9 | 4 |
| It is likely that my **institution** will encourage the | | | |
| implementation of goals ... after NSF grant expires. | 4.8 | 5.4 | 4 |

Generally, faculty reported that the project had a positive effect on their classroom instruction, their understanding of students, and their view of themselves. Women FF's tended to report a greater effect than the men faculty, but the differences are not considerable. It is interesting to note that neither the men or women respondents reported that they were encouraged to participate in this project by their department. This is consistent with their reports in focus groups that, generally, they did not believe the project goals would be furthered by their department after the funding was ended.

The findings of the FF surveys were consistent with information from the interviews and focus groups conducted during the project. Faculty reported that departments were less likely to be supportive of projects which were focused on classroom issues in general and gender issues in particular than those which focused on discipline specific research.

In May 1996 a second FF survey was sent to those who had participated in the project. The second survey was designed to collect information from those who were new to the project since the first survey and to get some sense of whether or not there were any changes in the responses of those who responded to the first survey. Of the 24 Faculty Fellows who were solicited, only nine returned the surveys and only 2 of those had not participated in first survey. Because of the small numbers, the analysis did not reveal any new information that could inform the project.

## General Findings

There were several important concepts that emerged as the project was implemented, and evaluated on a variety of campuses over time. These themes were "universal" over a variety of sites in this project and revolved around issues of faculty development and campus culture. These themes suggest lessons for other faculty or administrators who are interested in implementing new projects on their campuses.

- Institutional change takes time.
- Faculty development takes time and resources.
- Distinguished Visiting Professor-institutional fit is critical to project success.
- Faculty and administration "buy-in" are essential to institutionalization of project goals.
- Project activities must be validated by the faculty reward structure.
- Development of a "community" is essential to institutionalization of project goals.

***Institutional change takes time.*** Faculty and administrators on the local campuses recognized that changing the view of teaching and learning as it relates to gender issues was going to take time. In response to the question, "If there was one thing that you could recommend to the System Office to change in this program, what would you recommend?" one faculty member said, "Have [the DVP] stay here longer than a semester."

Because the project focused on the development of a small cadre of resident faculty to implement feminist pedagogy in the classroom and then work with other faculty, the impact of this project will be amplified over time. For most campuses, the short-term results were seen in the conversations among departmental faculty surrounding these issues and the impact on the faculty who were a part of the cadre. This was particularly true on campuses that had strong departmental support for the project. For example, when asked in a focus group, "What is THE most significant thing you learned from your participation in this project?" one faculty member said, "This program has changed the nature of conversations in my department . . . there are more discussions about methods, . . . this did not happen before." Others concurred.

Some of the FF's reported that resistance still existed in their departments and that some faculty were threatened by the perceived suggestion that their classroom behavior might be gender biased. This resistance seemed to be more pronounced on campuses where there was a lack of departmental support for the project.

***Faculty development takes time and resources.*** Clearly, institutional change takes time, but individual faculty development also takes time. At the final two sites, UW-Stevens Point and UW-La Crosse, Faculty Fellows were expected to become proficient as trainers of other faculty by the end of the semester that they worked with their Distinguished Visiting Professor. During their semester together, they worked with their DVP to develop a workshop that they could present to groups of faculty at other campuses. Faculty Fellows found this to be a problem. During a focus group with Faculty Fellows, the following comments were made:

> I am a tenure track person and I see these as golden opportunities, these are guaranteed presentation and publication opportunities for me . . . . I don't view them as hardships, but I do view [as a hardship] the fact that almost exclusively our Faculty Fellow time has been devoted to trying to plan them. So that has been the major disappointment for me.
>
> Maybe I am skipping ahead, but let me tell you what I think about why we have had some problems. It is time. I fully agree with the notion that the project is designed to produce trainers . . . . The notion is that we should be prepared to disseminate around the System. . . . I think my colleagues share my feelings that not having a good grip on what we have done differently to improve our classes, we would not want to tell anybody because we are not sure it even works yet.

Faculty development takes time *and* resources. Faculty Fellows consistently reported that it took time to understand and implement the teaching strategies in their own classrooms and or laboratories. In order for them to be successful, they felt the need to have the support of the department to provide some relief from heavy teaching loads. The project provided some support for FF release time, but it was often not given in the semester when the course or courses were being revised.

***Distinguished Visiting Professor–institutional fit is critical to project success.*** This project focused on bringing Distinguished Visiting Professors who were responsible for teaching a course in their area of expertise and providing faculty development for the FF's and other interested faculty. The campuses on which this strategy was the most successful were those where there was the best match between the background and experience of the DVP and those of the host campus faculty. Some of the most important factors were reported to be the type of institution the DVP came from, the match between the size and type of class they were teaching and that of the other faculty in the department they were visiting, and general personal characteristics. A Faculty Fellow in a focus group setting commented:

> You need to worry about a fit between the DVP and the Faculty Fellow or the institution or whatever I mean. And I don't know how you would do it. This hasn't been bad, but I can imagine situations that could have been God awful or I can imagine situations that could have been a whole lot better, but essentially you reach into a hat and the [start date of the term] somebody you have never met before arrives here. And how do you know the chemistry is going to work?

Another Faculty Fellow reflected on the issue of a good match between the DVP and the FFs in the following comments.

> Yeah, I think that I would have the same perception if [the DVP] was teaching the exact same size lecture as I am and if she was teaching the same lab, it would be easier to translate the advice of changes to make than it is when the lab manual has been rewritten effectively for what she is doing. . . . I don't imagine we are going to do that.

***Faculty and administration "buy-in" are essential to institutionalization of project goals.*** It was consistently reported that in order for the project to be successful and have any long-term effect, it was essential to have general faculty support. Good institutional match between Distinguished Visiting Professor and faculty contributed to faculty buy-in. This point is illustrated in the negative by a faculty member in a focus group setting who remarked that:

> I hope that we won't adopt some of those things because I have seen how very hard she [DVP] works with a much lighter schedule than I have. I can't imagine accomplishing those things, with the teaching loads we are given. I am tired already without attempting some of the labor-intensive things she does, not matter how good they may be.

On those campuses that reported having the greatest success, DVP's and FF's received strong administrative support. This support was manifested by positive feedback to the faculty involved in the project and explicit reinforcement of the goals of the project in their interaction with others on campus. In some cases, the campus provided monetary resources to continue the project in the absence of NSF funding. Project activities must be validated by the faculty reward structure. Faculty involved in the project consistently reported concern that their activity would be to their detriment in the promotion and tenure process. Almost one half of the FF's were not tenured and some did not perceive that there was support from their department for their involvement. The nature of the reward structure being weighted toward research and publication was seen as having a negative effect on the degree of involvement of other faculty. In commenting on the preparation of a workshop on pedagogy, a Faculty Fellow said:

> . . . that [workshop development] is not valued as research in our department. So it is like we have to do this plus some type of research in our discipline for tenure and promotion. So this is why we are talking at this meeting about either you -need release time or something because you can't take it [time to work on workshop development] out of your research time, at least not if you are going up for tenure, etc.

***Development of a "community" is essential to institutionalization of project goals.*** The overall success of the project was seen as being the result of the establishment of linkages within and among campuses, faculty working together, sharing experiences and creating a community of scholars. The support of others, faculty and administrators, was consistently reported as being critical to the faculty involved. Through the sharing of experiences, the faculty reported a new awareness of themselves, their students, and their discipline.

> I think the connection for me has been stronger on my campus, but I can see it building on other campuses, but not so much actually within my discipline. . . . Well, I think that the project has certainly given us the opportunity to get to know one another better . . . we know we're out there at the campuses where we can be reached. It is very reassuring.

## SUMMARY

Assessment and evaluation are central to systemic reform. They provide information for project developers to improve the project, evidence of the efficacy of the project, and needed information enabling key stakeholders to make determinations about institutionalization and dissemination of the project.

Early assessment and evaluation activities provided valuable feedback in shaping the nature of the project. The shift in emphasis from curricular development to faculty development was a response to information obtained from assessment activities. Improved administrative procedures and increased communication were also responses to formative assessment feedback.

The assessment and evaluation activities at all sites provided valuable lessons that may be helpful to other institutions in designing programs. Among those discussed in this report are:

- Institutional change takes time.
- Faculty development takes time and resources.
- Distinguished Visiting Professor–institutional fit is critical to project success.
- Faculty and administrative "buy-in" are essential to institutionalization of project goals.
- Project activities must be validated by the faculty reward structure.
- Development of a "community" is essential to institutionalization of project goals.

The assessment and evaluation activities of this project were designed to provide information to determine efficacy of the project and to provide information that could be a catalyst for reform activities. Evidence indicates that the assessment goals were met for the Collaborative Community. At the Collaborative Community, in partial response to assessment data, the project evolved, faculty members established learning communities that will help to institutionalize and disseminate the project, and institutional commitments to project goals have been made at some institutions, such as UW-River Falls. Evidence indicates that the overall goals of the project have been met at the Collaborative Community, and the Collaborative Community constitutes a national model for systemic reform.

# Appendix B

# Literature Review

**Nancy Mortell and Rebecca Armstrong**

**with Phyllis HolmanWeisbard and Laura Stempel**

This literature review presents suggested readings in a variety of topics covered in the preceding chapters. This is by no means a comprehensive list of all the available work on the general area of women and science, but the materials described here will give the reader a sense of its breadth, and most of the books and essays—as well as all of the Web sites—will point her in the direction of further reading. At the end of the review, a bibliography lists the main works consulted in *Flickering Clusters*.

## Women in Science

The experiences of women scientists—the barriers they faced, their career paths, and their accomplishments—has served as an important basis for feminist critiques of science. Evelyn Fox Keller's *A Feeling for the Organism: The Life and Work of Barbara McClintock* (1983) was a germinal work in this area and influenced many later studies of gender and science. This territory has also been explored in *Uneasy Careers and Intimate Lives: Women in Science* (1987), edited by Penina G. Abir-Am and Dorinda Outram, *The Outer Circle: Women in the Scientific Community* (1991), edited by Harriet Zuckerman et al., and by Margaret Rossiter in *Women Scientists in America before Affirmative Action*, 1940-71 (1995). Other books that examine the relationship between gender and science careers include *Lost Talent: Women in the Sciences* (1996), by Sandra L. Hanson; *Gender Differences in Science Careers: A Project Access Study* (1995), by Gerhart Sonnert with Gerald Holton, and *Who Succeeds in Science: The Gender Dimension* (1995), with life histories of ten men and ten women scientists, also by Gerhart Sonnert.

## Feminist Critique of Science

Numerous texts address the alienation of women from science, how science has constructed gender and framed how science is conducted and interpreted from an "objective" standpoint. Exemplary works in this critique of science include: Sandra Harding's *The Science Question in Feminism* (1986) and *Whose Science? Whose Knowledge?: Thinking from Women's Lives* (1991); Anne Fausto-Sterling's *Myths of Gender: Biological Theories about Women and Men* (1992); Ruth Bleier's *Feminist Approaches to Science* (1991) and *Science and Gender: A Critique of Biology and its Theories on Women* (1984); Jan Harding's *Perspectives on Gender and Science* (1986); Evelyn Fox-Keller's *Reflections on Gender and Science* (1985) and *Secrets of Life/Secrets of Death: Essays on Language, Gender, and Science* (1992); Sue Rosser's *Biology and Feminism: A Dynamic Interaction* (1992); Lynda I.A. Birke's *Women, Feminism and Biology: The Feminist Challenge* (1986); and Ruth Hubbard's *Profitable Promises: Essays on Women, Science and Health* (1995).

Other texts, such as *The Knowledge Explosion: Generations of Feminist Scholarship* (1992), edited by Cheris Kramarae and Dale Spender, *Science, Morality and Feminist Theory* (1987), edited by Marsha Hanen and Kai Nielsen, and *Feminism and Science* (1996), edited by Evelyn Fox Keller and Helen C. Longino, contain essays about feminism and specific science disciplines and/or integrate a feminist critique of human nature with modes of knowing. These theoretical works provide the basis from which a feminist perspective is brought to the classroom. Barbara Laslett's edited collection *Gender and Scientific Authority* (1996) reprints essays from the feminist journal *Signs*.

## Feminist Pedagogy/Education

In *Learning Our Way: Essays in Feminist Education* (1983), edited by Charlotte Bunch and Sandra Pollack, Charlotte Bunch's essay "Feminist Theory and Education" presents a model for teaching based on feminist theory in which the teacher describes what exists, analyzes why this reality exists, creates a vision of what should exist, and then provides strategies on how to make changes to what should be. In another essay, "Guidelines for a Teaching Methodology," Nancy Schniedewind builds on Bunch's model by providing practical methods of feminist teaching. In a more recent text, *Painful Pedagogy: Feminism in the Classroom* (1992), Lana Rakow examines the different teaching styles of men and women, how women professors negotiate classroom dynamics, and what constitutes a feminist teaching style. *Feminisms and Critical Pedagogy* (1992), edited by Carmen Luke and Jennifer Gore, contains essays reflecting on personal encounters with critical pedagogical practice and works toward formulating a viable feminist pedagogy of transformation.

In *The Feminist Classroom* (1994), Frances Maher and Mary Kay Thompson report on a research project which studied seventeen professors in eight different disciplines (including biology) to examine what feminist pedagogy is, and how knowledge is constructed in the classroom. Among their findings was a shifting web of relationships in the classroom—a community of learners—and the need for an awareness of the dynamics of race, gender, class, and sexual orientation informing all our constrictions. A collection of essays in *The Education Feminism Reader* (1994), edited by Lynda Stone, reveals how increased opportunities for some women conceal the discrimination that continues to exist in many private and public realms. The book examines issues of curriculum, knowledge construction, theorization of who teaches and how, and demonstrates the diversity of studies prevalent in education feminism today.

*Gender and Academe: Feminist Pedagogy and Politics* (1994), edited by Sara Munson Deats and Lagretta Tallent Lenker, offers an insightful look at the feminist influence on what is taught in colleges and universities, how it is taught and how feminists teach outside of the classroom. (Strategies such as mentoring are means by which feminists can facilitate change in the academy.) Written as one component of a three-year project, funded by a Sloane Foundation grant and conducted by the Association of Women in Science (AWIS), *A Hand Up: Women Mentoring in Science* (1993), edited by Deborah C. Fort, is intended to facilitate development and enhancement of activities designed to increase the number of women in science. The book is designed to serve as a source of support for women interested in pursuing careers in science and provides guidelines, resources and advice for both those seeking mentors and those willing to be mentors.

## Feminist Pedagogy Applied to Science Education

Eileen Byrne's *Women and Science: The Snark Syndrome* (1993) researches, reviews, and analyzes the commonly accepted beliefs and practices that form the basis for educational policies, emphasizing the fact that they are not founded on sound empirical research or substantive grounded theory. While this critique specifically addresses what the author refers to as "the snark syndrome" in Australia—where just because you say something three times, it's true—feminists in the United States can learn much from Byrne's critique and see parallels in their own institutions. Another international text that contributes to the examination of feminist pedagogy is *Feminism and Education: A Canadian Perspective* (1990), edited by Frieda Forman. This collection of essays addresses, among other issues, anti-racist feminist pedagogy, classroom discourse/practice, and curriculum content. In one essay on science education and self-confidence as a predictor of persistence in math and science, the author bridges the discourse on feminist pedagogy and science education, as well as noting the continuing problem of women's exclusion from and marginalization in the curriculum.

Feminist scientist-educators have taken this last issue to heart in recent years, and texts addressing pedagogy and curriculum issues have begun to appear.

Professor Sue Rosser, a Distinguished Visiting Professor in the UW Women and Science Project (and now a dean at Georgia Institute of Technology), has written several such texts. Rosser's feminist critique of science pedagogy and curriculum is most fully developed in the field of biology. Her first text in this area, *Teaching Science and Health from a Feminist Perspective: A Practical Guide* (1986), provides a theoretical context for a feminist perspective on science, discusses topics and issues in biology and health, and includes numerous examples of course syllabi and pedagogical suggestions for such courses. In *Female-Friendly Science: Applying Women's Studies Methods and Theories to Attract Students* (1990), Rosser summarizes the plethora of reports and critiques that speak to a crisis in higher education and shows how feminist theories and methods along with research by women scientists informed by women's studies scholarship can be applied to recruit and retain more women in science. As she notes, these methods also benefit people of color and men by making science a friendlier place for all students. Recognizing that most of the feminist critique of science and suggestions for improving the pedagogy and curriculum focus on the biological sciences, Rosser's edited collection *Teaching the Majority: Breaking the Gender Barrier in Science, Mathematics and Engineering* (1995) attempts to address and expand these same issues for other physical sciences, math, and engineering. Using Rosser's six phase model of curricular and pedagogical transformation to science, contributors to this book show how they have transformed their individual classrooms by modifying the curricula and their teaching techniques in the physical sciences with the intent of changing the composition and theoretical perspective of the pool of scientists.

While Rosser's text draws on many different disciplines, other texts, such as Joan Rothschild's *Teaching Technology from a Feminist Perspective: A Practical Guide* (1988) and Fran Davis and Arlene Steiger's *Feminist Pedagogy in the Physical Sciences* (1993), apply their strategies to a more limited field of study. *Mathematics and Gender* (1990), edited by Elizabeth Fennema and Gilah Leder, reviews the research on how men and women are treated differently in this discipline, which results in significant differences in achievement. They examine the learning environment and provide suggestions for classroom teaching. In *Women Changing Science: Voices from a Field in Transition* (1995), Mary Morse takes a current look at women scientists in society. Morse's work includes women doing science in all walks of life as well as a section of recommendations from her research on what could still be done to improve science for undergraduate women. This reference serves to remind us that in spite of the progress made to date, there is still much that needs to be done during the undergraduate years.

## Curriculum Transformation Viewed from Women's and Multicultural Perspectives

Although Women's Studies has been a recognized field of study for many years, its integration into the curriculum has been slow and almost nonexistent in some areas of the sciences. A number of texts address the issue of integrating women into the curriculum in general. The recommendations made by Elizabeth Higginbotham in *Integrating All Women into the Curriculum* (1988)—such as 1) increasing faculty knowledge with the new scholarship, especially with women of color, 2) developing ways to incorporate material without marginalization, and 3) creating a positive class environment for all student participation—speak to the goals and strategies of the UW project, which should be evident in the preceding narratives from faculty participants.

Curriculum transformation projects, such as those described by Betty Schmitz in *Integrating Scholarship by and about Women into the Curriculum* (1990) and in *Women of Color and the Multicultural Curriculum: Transforming the College Classroom* (1994), edited by Liza Filo-Matta and Mariam Chamberlain, include efforts at course revision, broader discipline paradigm shifts, the involvement of faculty, and barriers to change. In *Changing the Educational Landscape: Philosophy, Women and Curriculum* (1994), Jan Martin confronts what she describes as two of the dogmas of curriculum: God-given subjects and the immutable basics, courses and content that must be taught. In other cases, the barriers to change are not the faculty and the curriculum, but often the students themselves, highlighting the urgency with which students need to be exposed to, recognize, and learn to appreciate alternate and diverse perspectives. This not only applies to gender differences, but also to multicultural perspectives in education and science.

Using the approach of a matrix of domination, *Races, Class and Gender: An Anthology*, 2nd edition (1995), compiled by Margaret L. Anderson and Patricia Hill Collins, analyzes the interrelationship of race, class, and gender, and how these structures have shaped the experiences of all people in the United States. Focusing on the humanities and social sciences, *Race, Identity and Representation in Education* (1993), edited by Cameron McCarthy and Warren Crichlow, presents a broad view of what produces racial inequality in schooling and what cultural interventions are possible in school and society. *Blacks, Science and American Education* (1989), edited by Willie Pearson, Jr. and H. Kenneth Bechtel, sets the stage for an analysis of the status of African Americans in science and mathematics. Their comprehensive review of statistical data, coupled with case studies of intervention strategies aimed at students, highlights the importance of including blacks in any discussion of pedagogy, curriculum and climate in introductory science and mathematics courses.

Building on this earlier work, Willie Pearson, Jr., along with Alan Fechter, edited *Who Will Do Science?: Educating the Next Generation* (1994) which addresses the complexity of issues surrounding the recruitment and retention of blacks in science and mathematics. A particularly interesting chapter summarizes a research study on the intersection of black American culture and the organization and culture of academic science. They conclude that systemic reform informed by data collection and analysis with carefully crafted human resource policy is desperately needed. David Nelson, George Gherverghese Hoseph, and Julian Williams present the rationale for teaching from a multicultural perspective in *Multicultural Mathematics* (1993). Although focused on K-12 teaching, it provides examples of applications which can stimulate discussion amongst faculty members teaching introductory science and math courses.

## Reform of Science Education

With United States students' math and science proficiency scores dropping in recent years and fewer students entering the sciences, there has been much discussion within the disciplines about curriculum revision and innovation quite aside from any attempts to incorporate a feminist or multicultural perspective into the curriculum. Sheila Tobias' *They're Not Dumb, They're Different: Stalking the Second Tier* (1990) is probably one of the best known. In this work, Tobias attempts to explain why so many students who are qualified and capable of doing science choose another major in college. In *Revitalizing Undergraduate Science: Why Some Things Work and Most Don't* (1992), Tobias further examines the issue of retaining students in science and presents case studies of science programs that work in recruiting and retaining science majors by examining the programs, faculty, student, and institutional factors. R. D. Anderson's *Issues of Curriculum Reform in Science: Mathematics and Higher Order Thinking Across the Disciplines* (1994) focuses on K-12 curriculum reform, but specifically addresses the process of change in order for innovation to become the norm. Lynn Weber Cannon, in *Curriculum Transformation: Personal and Political* (1990), writes that curriculum transformation requires: 1) vision (a reconceptualization of one's discipline); 2) information (research and data on diversity of experience); and 3) a new pedagogy. The theme of future research is expanding to not only include curriculum and pedagogy evaluation, but also to be used to guide policy changes that are needed in higher education.

## Faculty Development

Sue Rosser and Bonnie Kelley offer a model for faculty development in *Educating Women for Success in Science and Mathematics* (1994), discussed at greater length in the body of the prospectus. In *Computer Equity in Math and Science: A*

*Trainer's Workshop Guide* (1991), Jo Shuchat and Mary McGinnis provide a sample of what might be possible when training faculty members to function as trainers to their colleagues in a specific discipline. Similarly, Laura Rendon's *Preparing Mexican Americans for Mathematics- and Science-based Fields: A Guide for Developing School and College Intervention Models* (1985) functions as a how-to guide for teachers, administrators, and parents of K-12 and college students and provides practical suggestions to encourage, retain and facilitate the success of Mexican-American students.

## Institutional Change

*The Equity Equation: Fostering the Advancement of Women in the Sciences. Mathematics and Engineering* (1996), written by Cinda-Sue Davis, Angela Ginorio, Carol Hollenshead, Barbara Lazarus, Paula Rayman and Associates, emerged from a conference sponsored by the Cross University Research in Engineering and Science group on gender issues in May 1994. This document reviews the current research literature and status of women in science, and suggests that solutions need to focus on institutional change. Arguing that future interventions and policy changes need to be grounded in research, they set out to define the research, policy and practice agenda in an effort "to promote broad public understanding of women's role in the sciences, mathematics and engineering" (327).

## Web Sites

Web sites dealing with women, minorities, and science have proliferated over the last few years, and this is just a selection of current ones. Each contains numerous links that will lead you to other relevant sites:

University of Wisconsin Women and Science Program:
*http://www.uwosh.edu/programs/wis*

University of Wisconsin System Women's Studies Librarian (includes links to other sites as well as access to bibliographies produced by the Librarian's office, some on women and science):
*http://www.library.wisc.edu/libraries/WomensStudies/*

Women and Minorities in Science (University of Illinois):
*http://www.physics.uiuc.edu/intemet/women.html*

Women and Minorities in Science and Engineering (an excellent collection of links from Ellen Spertus, MIT):
*http://www.ai.mit.edu/people/ellens/Gender/wom_and_min.html*

Iowa State University Program for Women in Science and Engineering:
*http://www.public.iastate.edu/-seema/pwse.html*

New Mexico Network for Women in Science and Engineering:
*http://www.ladmac.lanl.gov/nmnwse.html*

National Academy of Sciences Committee on Women in Science and Engineering:
*http://www.nas.edu/cwse/*

4000 Years of Women in Science (includes biographies, photos, and references):
*http://crux.astr.usa.edu/4000WS/4000WS.html*

National Center for Education Statistics (findings from 1997 study on women in math and science):
*http://www.ed.gov/NCES/pubs97/97982.html*

## BIBLIOGRAPHY


Abir-Am, P. G., and Outram, D. (Eds.) (1987). *Uneasy careers and intimate lives: Women in science, 1787-1979.* New Brunswick: Rutgers University Press.

Aikenhead, G. S., and Ryan, A. G. (1992). "The development of a new instrument: Views on science-technology-science," *Science Education* 75 (5): 477-491.

Andersen, M., and Collins, P. H. (1995). *Race, class, and gender: An anthology* (2nd ed.). Belmont, CA: Wadsworth.

Anderson, R. D. (1994). *Issues of curriculum reform in science mathematics and higher order thinking across the disciplines* (Contract no. RR91182001). Washington, DC: U. S. Government Printing Office.

Arambula-Greenfield, Teresa (1995). "Teaching science within a feminist pedagogical framework," *Feminist Teacher* 9 (3): 110-115.

Birke, L. I. A. (1986) *Women, feminism and biology: The feminist challenge.* Brighton: Wheatsheaf.

Bleier, R. (Ed.) (1986). *Feminist approaches to science*. New York: Teachers College Press.

Bleier, R. (1984). *Science and gender: A critique of biology and its theories on women*. New York: Pergamon Press.

Bleier, R. (1988). "A decade of feminist critiques in the natural sciences: An address by Ruth Bleier, with an introduction by J. W. Leavitt and L. Gordon," *Signs* 14 (1): 182-195.

Boyer, E. (199 ). *Scholarship redefined: Priorities of the professoriate*. Carnegie Foundation for the Advancement of Teaching.

Bruffee, K. A. (1992). "Science in a postmodern world," *Change* (September/October): 18-25

Bunch, C., and Pollack, S. (Eds.) (1983). *Learning our way: Essays in feminist education*. Trumansburg, NY: Crossing Press.

Byrne, E. M. (1993). *Women and science: The snark syndrome*. Washington, DC: Falmer Press.

Cannon, L. W. (1990). *Curriculum transformation: Personal and political*. Memphis: Memphis State University, Center for Research on Women.

Davis, C., Ginorio, A. B., Hollenshead, C. S., Lazarus, B. B., Rayman, P. M., and Associates (1995). *The equity equation: Fostering the advancement of women in the sciences, mathematics, and engineering*. San Francisco: Jossey-Bass.

Davis, F., and Steiger, A. (1993). *Feminist pedagogy in the physical sciences*. Montreal: Vanier College.

Deats, S. M., and Lendker, L. T. ( Eds.) (1994). *Gender and academe: Feminist pedagogy and politics*. Lanham, Md.: Rowman & Littlefield.

Fausto-Sterling, A. (1992). *Myths of gender: Biological theories about women and men* (2nd ed.). New York: BasicBooks.

Fennema, E., and Leder, G. C. (Eds.) (1990). *Mathematics and gender*. New York: Teachers College Press.

Filo-Matta, L., and Chamberlain, M. K. (Eds.) (1994). *Women of color and the multicultural curriculum: Transforming the college classroom*. New York: Feminist Press, City University of New York.

Forman, F. (Ed.) (1990). *Feminism and education: A Canadian perspective*. Toronto: Center for Women's Studies in Education, Ontario Institute for Studies in Education.

Fort, D. C. (Ed.) (1993). *A hand up: Women mentoring women in science*. Washington, DC: Association for Women in Science.

Greer, S. (1992). "Science: 'It's not just a white man's thing,'" *Winds of Change* 7 (2)

Hanen, M., and Nielsen, K. (1987). *Science, moral and feminist theory*. Calgary, Alta.: University of Calgary Press.

Hanson, S. L. (1996). *Lost talent: Women in the scientific community.* Philadelphia: Temple University Press.

Harding, J. (1986). *Perspectives on gender and science.* Philadelphia: Falmer Press.

Harding, S. G. (1986). *The science question in feminism.* Ithaca, NY: Cornell University Press.

Harding, S. G. (1991). *Whose science? Whose knowledge?: Thinking from women's lives.* Ithaca, NY: Cornell University Press.

Harding, S. G. (1995). *The "racial" economy of science.* Bloomington: Indiana University Press.

Higginbotham, E. (1988). *Integrating all women into the curriculum.* Memphis, TN: Center for Research on Women, Memphis State University.

Hubbard, R. (1995). *Profitable promises: Essays on women, science, and health.* Monroe, ME: Common Courage Press.

Keller, E. F. (1983). *A feeling for the organism: The life and work of Barbara McClintock.* W. H. Freeman.

Keller, E. F. (1985). *Reflections on gender and science.* New Haven, CT: Yale University Press.

Keller, E. F. (1992). *Secrets of life/secrets of death: Essays on language, gender, and science.* New York: Routledge.

Keller, E. F., and Longino, H. C. (Eds.) (1996). *Feminism and science.* Oxford: Oxford University Press.

Kramarae, C., and Spender, D. (Eds.) (1992). *The knowledge explosion: Generations of feminist scholarship.* New York: Teachers College Press.

Laslett, B. (Ed.) (1996). *Gender and scientific authority.* Chicago: University of Chicago Press.

Leiwan, S. (1994). "Four teacher-friendly postulates for thriving in a sea of change," *The Mathematics Teacher* 87 (8): 392-393.

Lou, R. (1994). "Teaching all students equally, in Teaching from a multicultural perspective," *Survival skills for scholars, vol. 12.* Thousand Oaks, CA: Sage Publications.

Loving C. C. (1991). "The scientific theory profile: A philosophy of science model for science teachers," *Journal of Research in Science Teaching* 28 (9): 823-838.

Luke, C. and Gore, J. (Eds.) (1992). *Feminisms and critical pedagogy.* New York: Routledge.

Lyons, N. (1994). "Dimensions of knowing: Ethical and epistemological dimensions of teachers' work and development," in Stone, L. (Ed.) (1994). *The education feminism reader.* New York: Routledge.

Maher, F. A., and Thompson, M. K. (1994). *The feminist classroom.* New York: Basic Books.

Malcolm, S. M. (1996). "Science and diversity: A compelling national interest," *Science* 271: 1817.

Malcolm, S. M., Aldrich, M., Hall, P. W., Boulware, P., and Stern, V. (1984). *Equity and excellence: Compatible goals*. Washington, D.C.: American Association for the Advancement of Science.

Martin, J. R. (1994). *Changing the educational landscape: Philosophy, women, and curriculum*. New York: Routledge.

Matthews, M. (1994). *Teaching science: The role of history and philosophy of science*. New York: Routledge.

McCarthy, C., and Crichlow, W. (1993). *Race, identity, and representation in education*. New York: Routledge.

Middlecamp, C., and Baldwin, O. (1995). "The Native American Indian student in the science classroom: Cultural clash or match?," in *Proceedings of the Third International History, Philosophy, and Science Teaching Conference*.

Miller, S. K. (1992). "Asian-Americans go bump against glass ceilings," *Science* 258 (5085): 1224.

Morse, M. (1995). *Women changing science: Voices from a field in transition*. New York: Insight Books.

Nelson, D., Joseph, G. G., and Williams, J. (1993). *Multicultural mathematics*. New York: Oxford University Press.

Nelson-Barber, S., and Estrin, E. T. (1995). "Bringing Native American perspectives to mathematics and science teaching," *Theory into Practice* 34 (3):

Orr, E. W. (1987). *Twice as less*. New York: W. W. Norton.

Pearson, W., and Bechtel, H. K. (Ed.) (1989). *Blacks, science, and American education*. New Brunswick: Rutgers University Press.

Pearson, W., and Fechter, A. (Ed.) (1994). *Who will do science? Educating the next generation*. Baltimore: John Hopkins University Press.

Rakow, L. F. (1992). *Painful pedagogy: Feminism in the classroom*. New Orleans, LA: Newcomb College of Tulane University.

Rendon, L. I. (1985). *Preparing Mexican Americans for mathematics- and science-based fields: A guide for developing school and college intervention models*. Educational Resources Information Center (ERIC) Clearinghouse on Rural Education and Small Schools (CRESS), New Mexico State University.

Rosser, S. V. (1986). *Teaching science and health from a feminist perspective: A practical guide*. New York: Pergamon Press.

Rosser, S. V. (1989). "Ruth Bleier: A passionate vision for feminism and science," *Women's Studies International Forum* 12 (3): 249-252.

Rosser, S. V. (1990). *Female-friendly science: Applying women's studies methods and theories to attract students*. New York: Pergamon Press.

Rosser, S. V. (1992). *Biology and feminism: A dynamic interaction*. New York: Twayne Publishers.

Rosser, S. V., and Kelley, B. (1994). *Educating women for success in science and mathematics*. West Columbia, SC: Wentworth Printing.

Rosser, S. V. (Ed.). (1995). *Teaching the majority: Breaking the gender barrier in science, mathematics, and engineering*. New York: Teacher College Press.

Rosser, S. V. (1997). *Re-engineering female friendly science*. New York: Teachers College Press.

Rossiter, M. (1995). *Women scientists in America before affirmative action, 1940-72*. Baltimore: Johns Hopkins University Press.

Rothschild, J. (1988). *Teaching technology from a feminist perspective: A practical guide*. New York: Pergamon Press.

Sanders, J. S., and McGinnis, M. (1991). *Computer equity in math and science: A trainer's workshop guide*. Women's Action Alliance. Metuchen, NJ: Scarecrow Press.

Sandler, B. R., Silverberg, L. A., and Hall, R. M. (1996). *The chilly classroom climate: A guide to improve the education of women*. Washington, D.C.: National Association for Women in Education.

Schmitz, B. (1990). *Integrating scholarship by and about women into the curriculum*. Memphis, TN: Center for Research on Women, University of Tennessee.

Seymour, E., and Hewitt, N. M. (1994). *Talking about leaving: Factors contributing to high attrition rates among science mathematics and engineering undergraduate majors*. Boulder, CO: Bureau of Sociological Research, University of Colorado.

Shuchat, J., and McGinnis, M. (1991). *Computer equity in math and science: A trainer's workshop guide*. Women's Action Alliance. Metuchen, NJ: Scarecrow Press.

Sonnert, G. (1995). *Who succeeds in science: The gender dimension*. New Brunswick: Rutgers University Press.

Sonnert, G., with Holton, G. (1995). *Gender differences in science careers: The Project Access study*. New Brunswick: Rutgers University Press.

Stone, L. (Ed.) (1994). *The education feminism reader*. New York: Routledge.

Tobias, S. (1990). *They're not dumb, they're different: Stalking the second tier*. Tucson, AZ: Research Corporation.

Tobias, S. (1992). *Revitalizing undergraduate science: Why some things work and most don't*. Tucson, AZ: Research Corporation.

Tobin, K. (Ed.) (1995). *The practice of constructivism in science education*. Hillsdale, NJ: Lawrence Erlbaum.

Whatley, M. H. (1989). "A feeling for science: Female students and biology texts," *Women's Studies International Forum* 12 (3): 355-361.